A Computational Model of Dissipation of Oxygen from an Outward Leak of a Closed-Circuit Breathing Device

Kathryn M. Butler
Building and Fire Research Laboratory
Fire Research Division

June 2007

U.S. Department of Health and Human Services
Mike Leavitt, Secretary

Centers for Disease Control and Prevention
Julie Louise Gerberding, Director

National Institute for Occupational Safety and Health
John Howard, Director

U.S. Department of Commerce
Carlos M. Gutierrez, Secretary

Technology Administration
Robert Cresanti, Under Secretary for Technology

National Institute of Standards and Technology
William Jeffrey, Director

ABSTRACT

Closed-circuit breathing devices recycle exhaled air after scrubbing carbon dioxide and adding make-up oxygen from a tank of pure oxygen. Use of this equipment allows first responders to work for up to four hours without swapping out cylinders and scrubbing canisters. Firefighting situations in which these devices would be useful include tunnels, mines, ships, high-rise buildings, and environments contaminated with biological or chemical toxins. A risk perceived by firefighters entering environments containing open flame and high radiant heat is the possibility of fire ignition in the vicinity of the respirator caused by the outward leakage of oxygen around the facepiece.

This paper presents the results of a computational fluid dynamics (CFD) study of oxygen dissipation into the environment surrounding a respirator facepiece. Actual heads and masks are scanned into a 3D data set for entry into the CFD software, providing a physical boundary for the problem to be solved. Leak geometries representing an imperfect seal are defined. Oxygen concentration fields and flow streamlines are presented for multiple combinations of fuel and air in the surrounding environment, for pure oxygen and air expelled from the leak, and for both normal and high stress breathing patterns. The flammability diagram for propane is used to estimate the flammable regions as a function of time during two breathing cycles for each case.

ACKNOWLEDGEMENTS

The preparation of this report was supported by the Office of Law Enforcement Standards under Interagency Agreement # HSSCHQ-04-X-00641, and by the National Institute for Occupational Safety and Health (NIOSH) National Personal Protective Technology Laboratory (NPPTL) as part of the project entitled "Task for Computer Simulation of Face Piece Oxygen Leakage Under Positive Pressure". The results published herein are solely the responsibility of the author. The author would like to thank John Kovac of NPPTL for his oversight on this project and many helpful discussions. Nicholas Kyriazi of NPPTL helped us understand the experimental evidence. Informative discussions on respirators with Ronald Shaffer, Ziqing Zhuang, and William Newcomb of NPPTL are also gratefully acknowledged. Daniel Crowl of Michigan Technological University provided valuable information on flammability diagrams. Dennis Viscusi of NIOSH provided the 3D scanned image of the headform used in NPPTL experments, and Dräger Safety provided technical drawings for a respirator. Technical discussions with Rodney Bryant and Nelson Bryner of NIST are gratefully acknowledged.

ABSTRACT I

ACKNOWLEDGEMENTS II

1   INTRODUCTION 1

2   MODEL CONDITIONS 3

2.1   Breathing Rates...3

2.2   Leaks 4

2.3   External Environment 5

3   MODEL GEOMETRY 7

3.1   Headform 7

3.2   Respirator Geometry 8

3.3   Combining Headform and Respirator..9

4   COMPUTATIONAL FLUID DYNAMICS MODELING 12

4.1   Flow..12

4.2   Chemistry...13

4.3   Turbulence 14

4.4   Problem Geometry 15

4.5   Boundary Conditions..16

4.6   Initial Conditions 18

5   MESH STUDY 19

5.1   Leak resolution..19

5.2   Size of External Box..21

5.3   Mesh Refinement in Direction of Jet 23

6   RESULTS FOR A SMALL LEAK DURING BREATHING AT REST 26

6.1   Leak of Pure Oxygen into Pure Propane Environment.......................26

6.2   Leak of Pure Oxygen into 10 % Propane Environment.......................30

6.3   Leak of Air into 10 % Propane Environment..32

6.4   Leak of Pure Oxygen into 5 % Propane Environment.........................33

7   RESULTS FOR A SMALL LEAK WHILE BREATHING UNDER STRESS 37

7.1   Leak of Pure Oxygen into 10 % Propane Environment.......................37

8   CONCLUDING REMARKS 39

REFERENCES 40

1 INTRODUCTION

The respiratory system provides a major access route into the body for chemicals, particulates, and bioactive agents. Respirators protect the body by preventing hazardous gases, vapors, and particles from entering the respiratory system.

Atmosphere-supplying respirators, which incorporate a self-contained air supply, are used in environments that are considered immediately dangerous to life or health (IDLH). These environments may contain highly toxic respiratory contaminants or too little oxygen to sustain life. While fighting a fire or overhauling a building, firefighters often work in environments where the respiratory hazards are unknown. Firefighters are therefore provided with self-contained breathing apparatus (SCBA) units, with a clean air supply cylinder that is carried on the wearer's back. These respirators also provide the unrestricted mobility required during firefighting.

Open-circuit SCBAs, which vent exhaled air to the atmosphere, are by far the most commonly used in firefighting. The standard compressed air cylinders used in these devices contain a limited air supply with a duration of 30 min to 60 min when breathing rates are normal. Under heavy workloads, a compressed air cylinder may last as little as 10 min to 20 min. Although this may be adequate for fighting ordinary house fires, longer durations may be necessary in certain situations, including fires in tunnels, mines, ships, and high-rise buildings. A situation of particular concern after the events of 9/11 is that of a terrorist attack, in which a fire is accompanied by possible chemical, biological, radiological or nuclear (CBRN) contamination.

An alternative to the open-circuit SCBA is the closed-circuit Self-Contained Breathing Apparatus (CC-SCBA), which recirculates exhaled air by absorbing carbon dioxide and adding fresh oxygen. Because oxygen needs are only around 5 % of air needs (oxygen consumption rate/ventilation rate),[1] an oxygen cylinder of comparable weight to a compressed air cylinder can sustain the firefighter for a much longer period of time. Closed-circuit SCBAs, also known as rebreathers or four-hour sets, can be used for up to 4 h before swapping cylinders and CO_2-absorbent canisters. This type of equipment is currently used as rescue breathing apparatus in mining operations.[2,3]

The National Institute for Occupational Safety and Health (NIOSH) is developing standards for the use of closed-circuit SCBA by firefighters and other first responders after a terrorist attack.[4,5] In addition to the respiratory protection of first responders in environments containing CBRN agents identified as inhalation hazards, the equipment must be usable under conditions of high heat and open flames. Although a leak is unlikely for a respirator with good fit, the possibility of outward leakage of oxygen in a fire environment is a concern. If the oxygen remains concentrated around the face in the presence of fuel gases and heat, ignition may occur. If the oxygen diffuses rapidly away from the face, however, the possibility of ignition is negligible. The intent of this study is to consider the potential for danger under worst-case conditions.

Computational fluid dynamics (CFD) numerically simulates fluid flow by solving the equations of motion. The solutions can be obtained in detail that is impossible to achieve in experiments, and for variables that are difficult or impossible to measure in practice, especially in three-dimensional space. A variety of situations, including breathing pattern, leak geometries, and varying concentrations of gaseous components, may be tested. The visualization of the computational results as they vary with time and space enhances our understanding of the flow phenomena. Validation of CFD with experimental measurements is recommended to ensure that the results accurately reflect reality.

2 MODEL CONDITIONS

In this report, respirator leakage effects on the environment exterior to the mask are investigated over a set of conditions that are within the range expected to be encountered by a firefighter. The variables of primary interest in this study are breathing rate, leak geometry, and external environment.

2.1 Breathing Rates

The pulmonary system is capable of supporting a variety of activity levels. Increased workloads require increased oxygen to the bloodstream, which is obtained by a combination of increased breathing rate and increased volume of air per breath. The respiratory minute volume or minute ventilation $\dot{V}_E$ is defined as the volume of air that is moved out of a person's lungs in one minute, and is given by the product of the tidal volume V_T, the volume of air moved during the exhalation phase of each breath, and the respiratory frequency f, the number of breaths per minute. For an average adult, values of $\dot{V}_E$ can range from about six liters per minute at rest to a maximum value of (70 to 90) L/min during strenuous exercise. For well-trained or elite athletes, maximum $\dot{V}_E$ can easily exceed 100 L/min. In a study of firefighters, the median minute volume during heavy work rates was found to be about 60 L/min, with a capability of 100 L/min or more over short time periods.[6] So the minute ventilation under a heavy workload may be 12 to 15 times that at rest or even more.[7]

While at rest, an average adult breathes at a rate of (12 to 15) breaths per minute, with a tidal volume of about 0.5 L. (The maximum capacity of the lungs is about ten times this volume.) Inhalation requires the contraction of the diaphragm, and the flow rate rises and falls more or less symmetrically in time, with variability due to changes in the rate of contraction. During normal exhalation the diaphragm relaxes, and the flow rate quickly rises to a maximum value then decreases in time. Durations of inhalation and exhalation cycles may not be exactly the same.

For the respirator leak model studied here, a simplified breathing cycle is assumed. For a normal resting breath, the respiratory frequency is taken to be 15 breaths per minute, with inhalation and exhalation cycles each lasting two seconds. The flow during exhalation is considered to be steady over time, with a tidal volume of 0.5 L released over the 2 s duration, a fraction of which is expelled steadily through the leak. During inhalation, the leak is assumed to close. This assumes that the low positive pressure that exists within the respirator facepiece during inhalation is insufficient to keep the leak open.

This study also considers a breathing cycle at high breathing rate and tidal volume, representing breathing under high stress due to exertion. This cycle takes the same form as the breath at rest, with a breathing rate of 60 breaths per minute and tidal volume of 3 L. This gives a minute ventilation of 180 L/min, which is very high.

2.2 Leaks

Most CC-SCBAs are positive pressure respirators, which maintain positive pressure in the facepiece throughout the breathing cycle. An outward leak is therefore a possibility if a perfect seal between respirator and face is lost.

In order to use a respirator with a tight-fitting facepiece, a firefighter must undergo fit testing using the same make, model, style, and size of respirator that will be worn on the job. Fit testing must be performed at least once a year, whenever there is a change in the respirator, and whenever there is a change in the physical condition of the face, such as dental changes and weight gain or loss. The fit can also be affected by facial hair and breathing demands caused by high work rates. A seal check performed by the user every time the respirator is put on helps to ensure a tight seal.

The most likely position and size for an outward leak from a respirator facepiece is not known. In the absence of this information, a leak location near the temples was selected for this study. The size of the leak was arbitrarily chosen to be 1 mm wide by 43.6 mm long – larger than a pinhole leak but not catastrophic in extent.

The mixture of gases expelled by a leak is determined by the gaseous content of the facepiece. In normal breathing without a respirator, an exhalation contains approximately 16 % O_2 by volume and 4 % CO_2 by volume. When using the CC-SCBA, the exhaled air passes through a canister that absorbs the CO_2, then into a flexible breathing bag. For most types of apparatus, the O_2 is supplied to the breathing bag at a steady rate from the oxygen source, which can be either compressed O_2 gas, liquid O_2, or solid potassium superoxide (KO_2). With compressed-O_2 apparatus, when O_2 consumption is greater than the O_2 flow rate, a demand valve supplies more O_2 as needed. When the breathing bag is full, a relief valve vents excess contents to the ambient environment.

The concentration of oxygen within the respirator facepiece must therefore be somewhere between the 16 % of normal exhalation and the 100 % content of the oxygen source. (Note that below 19.5 % a gas mixture is considered to be oxygen deficient.) Tests performed by NIOSH on fourteen rescue breathing apparatus showed that, for all but two units, the average inhaled oxygen concentration was much higher than that of the exhaled breath.[2] In fact, average O_2 concentrations were above 60 % for eleven units and above 80 % for six.

The danger posed by an outward leak into a mixture that contains flammable gases depends, among other things, on the concentration of oxygen in the facepiece. Since the intent of this study is to investigate a worst-case scenario, and since elevated levels of oxygen have been measured within the facepieces of several CC-SCBA units during operation, the model will assume that the content of the gas expelled through the leak is pure oxygen.

2.3 External Environment

The incomplete combustion that takes place in an uncontrolled fire generates smoke and flammable fuel gases that may be transported a distance from the fire. One of the hazards faced by the firefighter, and a primary reason for using an atmosphere-supplying respirator, is a lack of knowledge about the contents of the surrounding atmosphere. Especially while fighting a fire in the same room, a firefighter may be subjected to high concentrations of flammable gases. It is important to understand what the possible consequences are if oxygen from a respirator mask leak is introduced into this environment.

For this study, a way is needed to estimate the potential for fire to occur in the vicinity of a leak of oxygen into a fuel/air mixture. First, consider a mixture of a fuel gas with air. A mixture of fuel and air is able to burn only when the concentration of the fuel gas is within a certain range of values, delineated by the lower flammable limit, or LFL, and the upper flammable limit, or UFL. Below the lower flammable limit, the mixture of fuel and air lacks sufficient fuel to burn. Above the upper flammable limit, the mixture is too fuel rich and lacks sufficient oxygen to burn. Flammable limits have been determined experimentally for many gases and vapors.[8] The ignition of a mixture within the flammable range requires a small source of energy, which can be provided, for example, by a spark due to static electricity.

In the problem described in this paper, pure oxygen is introduced to a fuel/air mixture. An estimate of the potential for fire to occur in the vicinity of a leak of oxygen into a fuel/air mixture requires a flammability diagram for the fuel.[9,10] For this project, propane is chosen as a simple gaseous hydrocarbon fuel.

Fig. 1 shows a flammability diagram for propane. In this diagram, concentrations of propane, oxygen, and nitrogen (fuel, oxidant, and inert gas respectively) are shown on three axes. Each point within the diagram describes a unique mixture of the three gases. Concentrations are given in percent by volume. Combinations of air with fuel are found along the air/fuel line, which connects pure propane at point D at the top, to pure air at point E at the bottom (79 % nitrogen, 21 % oxygen, and 0 % propane). The LFL and UFL for propane in air are located along this line at values of 2.1 % by volume and 9.5 % by volume.[8] These are plotted at points B and A respectively. The line from point D that is tangent to the flammable region, at point C, is known as the limit line. The mixtures along the limit line have a fixed ratio of oxygen to nitrogen below which flame propagation is not possible for any amount of propane. This line meets the lower axis at the limiting oxygen concentration, LOC, which is 11.5 % by volume for propane.[10]

In pure oxygen, the LFL and UFL have been determined experimentally for several fuels. Values for propane were not found in the literature, but a reasonable estimate can be obtained using values for other simple hydrocarbon fuels. The UFL in oxygen is 61 %, 66 %, 49 % and 52 % for methane, ethane, n-butane, and n-hexane respectively.[11] These are simple linear hydrocarbons with one, two, four, and six carbon atoms respectively.

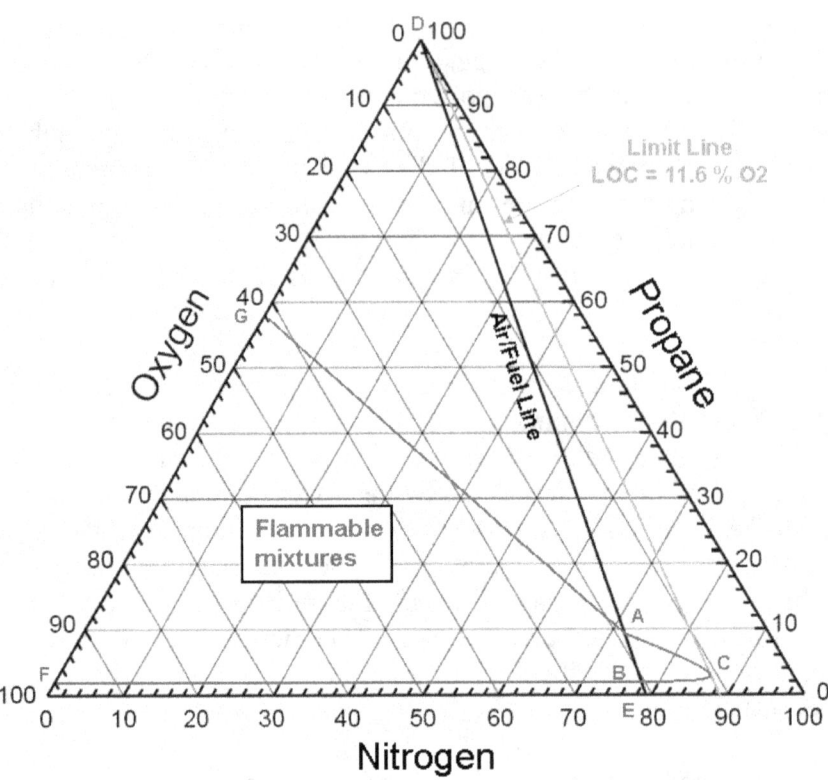

Figure 1. Flammability diagram for propane (in volume percent). UFL for oxygen (Point G) is estimated.

According to the principles of the group contribution method,[12] molecules are expected to behave according to the number and type of functional groups within them. Thus the UFL in oxygen of propane, with three carbon atoms, can be estimated as about 58 % by volume by plotting UFL versus number of carbons. For most fuels, the LFL in oxygen is not much different than the LFL in air. The LFL in oxygen of propane is therefore estimated as 2.1 % by volume, the same as that in air. The LFL and UFL in oxygen are plotted in Fig. 1 as points F and G respectively, along the axis with zero nitrogen concentration.

The connection of points F, B, C, A, G in the flammability diagram provide an estimate of the flammable region for propane.[13] All mixtures within this region can be ignited. Mixtures outside of this range cannot.

The flammability diagram is a function of temperature and pressure. For the values presented here, the conditions are 25° C and atmospheric pressure.

3 MODEL GEOMETRY

Experiments using mannequins have demonstrated that a simplified cylindrical geometry is not adequate to represent the flow field near a human face.[14] The complex features of the human face combined with the features of a respirator mask are necessary for an accurate flow model.

For this modeling effort, the complex geometry for a human head wearing a respirator facepiece has been obtained by assembling the shapes for a headform and a facepiece that were developed separately.

3.1 Headform

A headform used by the National Institute for Occupational Safety and Health (NIOSH) in respirator breathing experiments was digitized for use in this model. NIOSH used a 3D head scanner to capture the three-dimensional external profile of the headform in the form of a point cloud, which is a large set of data points with values in x, y, and z directions. These points define the location of the surface. The scanner software arranged this set of points into an IGES file, a standard format for Computer Aided Design (CAD) systems, which was then provided by NIOSH to the National Institute of Standards and Technology (NIST).

A software package called Raindrop Geomagic was used to convert the point cloud to a set of geometrical surfaces. Fig. 2 illustrates the development of the headform geometry. In Fig. 2(a), the point cloud is shown, with over 180 000 points. The object sticking out of the back of the headform is a holder used to keep it in place during scanning. Points representing the holder were removed first. Fig. 2(b) shows an early stage of model development after the point cloud was converted into a set of triangles. At this point there were some holes in the geometry; in particular at the mouth, where the physical headform has an opening, and on the top of the head, where the scanner was unable to see and therefore could not fix any points. The ridge that surrounds the face is clay that is molded onto full facepiece masks to provide a seal during experiments. The Geomagic software has a variety of tools for filling holes and smoothing surfaces. Fig. 2(c) shows the headform geometry after it was cleaned up. The final step in Geomagic was the conversion of the set of triangles to a few hundred patches, or 3D surface entities, that were then exported to another IGES file as NURBS surface data.

The computational fluid dynamics (CFD) software package used for this project, CFD-ACE+, has a pre-processor, CFD-GEOM, for geometry development. The IGES file from Geomagic was read into CFD-GEOM, resulting in a set of surface entities that describe the 3D surface of the headform. The surfaces can be changed to develop the geometry further.[15] In Fig. 2(d), the boundaries of surfaces around the face were smoothed to eliminate the clay ridge. The final step in preparing the headform geometry was to split it into two along a vertical plane of symmetry. As is discussed below, the use

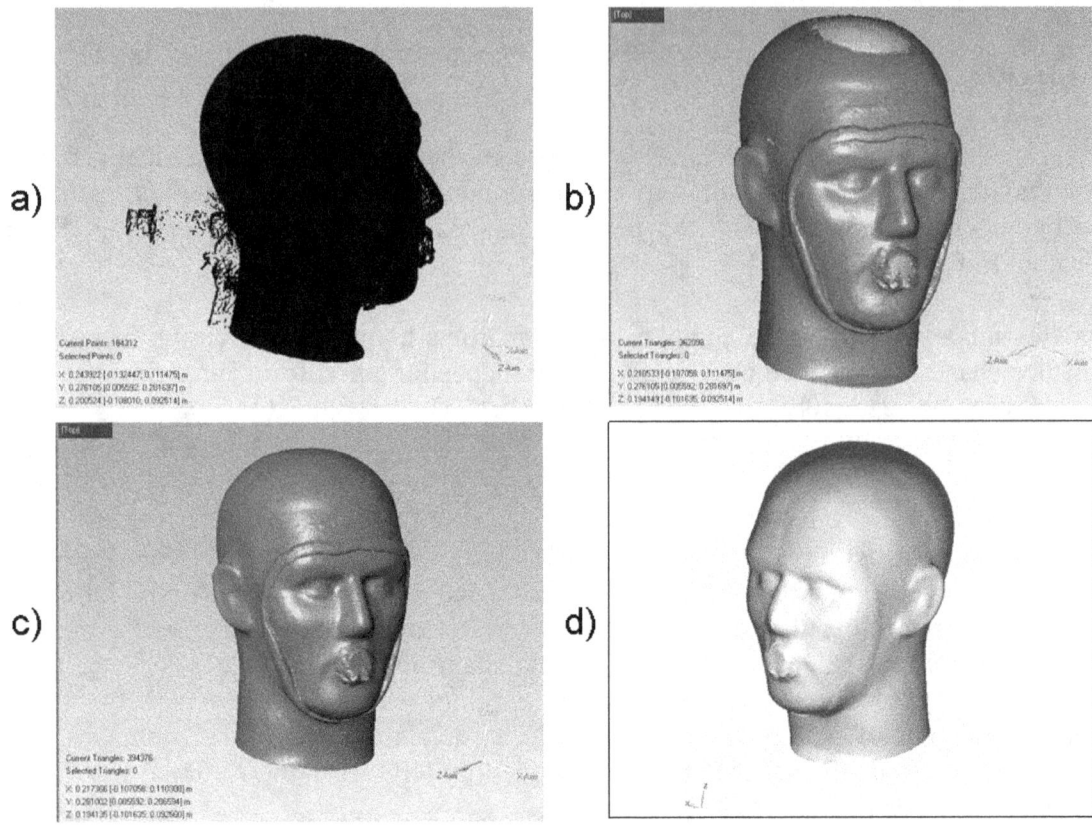

Figure 2. Steps in preparing headform geometry: a) point cloud from 3-D scanner, b) conversion into triangles, c) holes filled in, d) clay removed.

of symmetry allowed the problem size to be halved. To accomplish this final step, all surface entities that intersected the symmetry plane were re-defined such that one edge was along the plane.

3.2 Respirator Geometry

For a prototypical full facepiece respirator, mechanical drawings were obtained from the manufacturer. This allowed a geometry to be built in CFD-GEOM from the relative locations of points, curves, and surfaces shown in multiple views and cross-sections. Figs. 3(a) and (b) show front and side views of points, lines and curves that were used during construction. Figs. 3(c) and (d) show two views of the completed facepiece. In these latter two illustrations the visor and facepiece connector region below the visor have been left open to allow better viewing of the geometry. For final preparation of the model, both visor and facepiece connector are closed. Note that the inner mask, which covers the mouth and nose, is missing in this model. Since this project is concerned only with leaks to the outside environment, the inner mask is not needed.

8

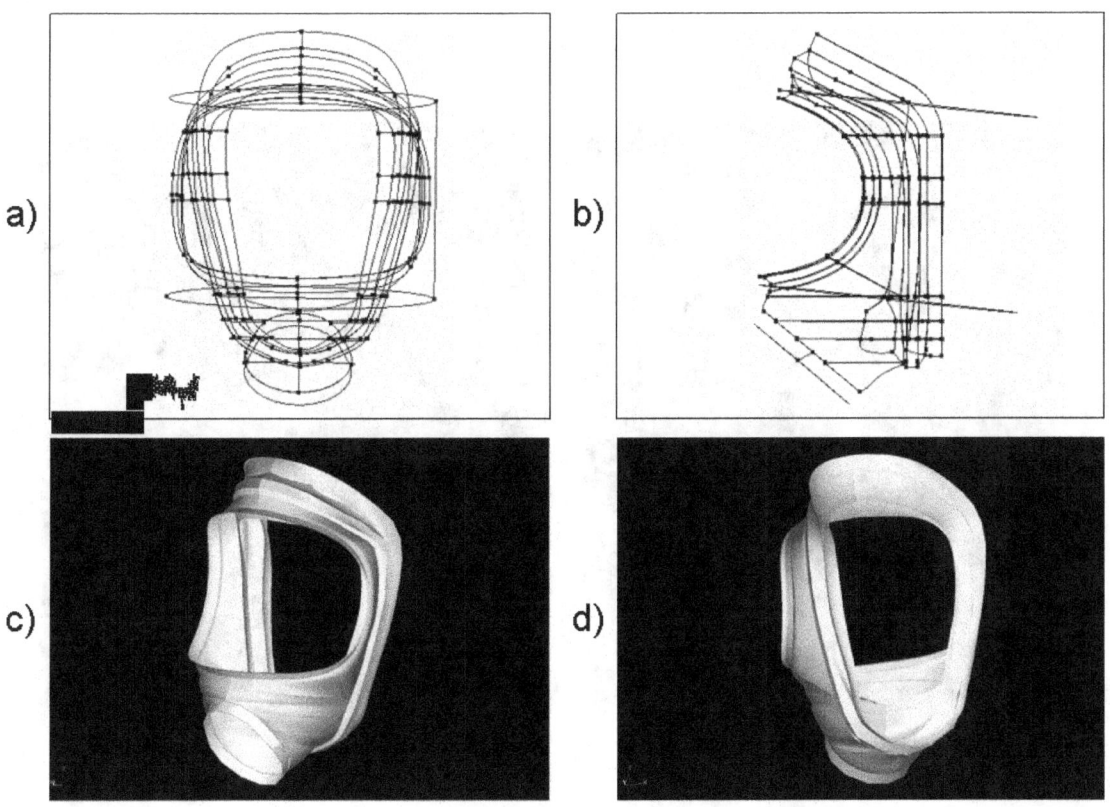

Figure 3. Preparation of mask geometry: a) and b) are wire frame images of 3-D geometry derived from mechanical drawings, and c) and d) are perspective views.

Only half of the facepiece was actually constructed from the mechanical drawings. The other half was obtained by mirroring the points and curves about the plane of symmetry. Halving the facepiece to take advantage of symmetry, therefore, was a trivial operation.

3.3   Combining Headform and Respirator

To combine the head with the facepiece, the headform was translated and rotated into the proper position relative to the mask within CFD-GEOM, with planes of symmetry oriented in the same direction. The first attempt, simply merging the two geometries, did not provide a good fit. With the chin in good position, the top of the facepiece was not in contact with the forehead, as shown in Fig. 4(a), and the sides extended inside of the headform by more than a centimeter. In the real world this geometrical mismatch is eliminated by the flexibility of the facepiece, which allows it to fit the shape of the head as it is put on. A good fit was achieved in the virtual world through adjustment of the facepiece geometry by moving the top toward the forehead, pulling out the sides, and defining the inner and outer seals to follow the contours of the face, as shown in Fig. 4(b). In the real world, surface areas of sections of the respirator remain constant while the angles between them adjust to fit the head.   Maintaining these physical

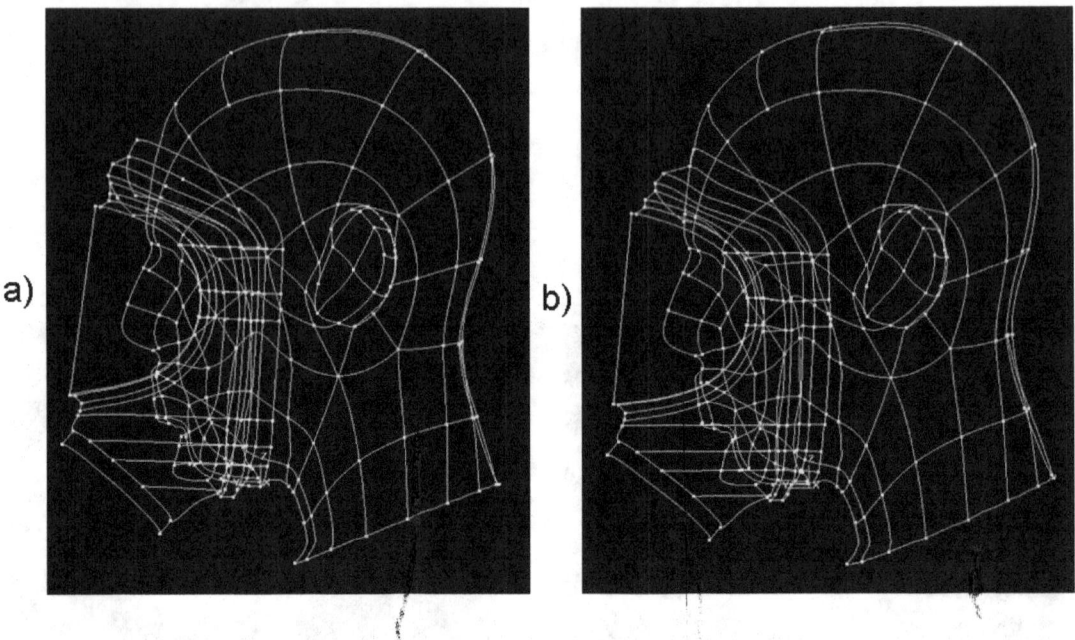

Figure 4. Side view showing fitting of respirator mask to headform: a) original 3-D geometries of mask (yellow) and headform (blue) brought together, b) after adjustment of mask geometry.

relationships was not attempted, although an effort was made to prevent gross deformities in the shape of the facepiece.

The final step in making a single model out of two geometries was to redefine surface entities in contact with each other such that adjacent surfaces share a common boundary. Fig. 5 shows the combined headform and facepiece after processing, as they appear in CFD-GEOM.

In this study, the geometry for a leak is defined by the modeler. A representative leak was selected along the outer seal of the facepiece just above the temple region, where straps are attached. The leak is shown in Fig. 5 as a thin red line in front of the ears. It represents a thin breach of the outer seal about 1 mm thick and 43.6 mm long.

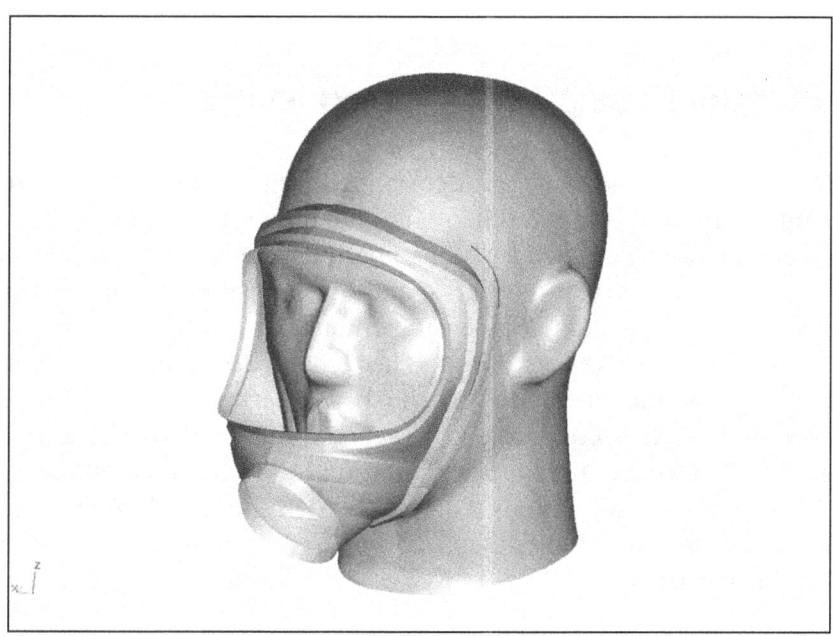

Figure 5. Model geometry for headform combined with respirator facepiece. Red mark indicates the location selected for a representative leak to be studied with this model.

4 COMPUTATIONAL FLUID DYNAMICS MODELING

The flow of gases and liquids can be determined using mathematical equations that describe the conservation of mass, momentum, and energy within the space occupied by the fluid. The equations are derived by balancing the time rate of change of the quantity of interest in a small volume with its flux in and out, transport due to diffusion, and any sources or sinks within the volume. The momentum conservation equation for fluids is also known as the Navier-Stokes equation. The terms that are required in the equations depend on the physics and chemistry of the particular problem to be solved. The equations provide only an approximation to reality – the challenge is to make sure that the most important factors are included. In addition to the fluid dynamics equations, the mathematical description of a problem requires the definition of boundary conditions on all boundaries defining the problem space. For a time-dependent problem, initial conditions must be specified.

The subfield of computational fluid dynamics (CFD) develops numerical methods and algorithms for the accurate and efficient solution of the equations for fluid flow. Many computational fluid dynamics codes have been written to solve various types of problems. The needs of a model of oxygen leaking from a respirator facepiece were found to be met by a commercial CFD software package called CFD-ACE+. CFD-ACE+ is a multiphysics code that is capable of solving a wide variety of problems, including fluids, heat transfer, chemistry, biochemistry, structural dynamics, electricity, and electromagnetism. In this application, flow, chemistry, and turbulence modules have been used.[16] The version of CFD-ACE+ used to develop and run this model was V2004.

4.1 Flow

The CFD-ACE+ software is based on the finite volume method. With this approach, the solution space is divided into a large number of cells, also known as control volumes. The cells are arbitrarily shaped triangles or quadrilaterals in two dimensions, and arbitrary 4- to 6-sided volumes in three dimensions. This allows the discretization of complex solution spaces, such as the geometry for the combined headform and facepiece in this model.

The governing equations to be solved are integrated numerically over each computational cell. The average cell value of all dependent variables (such as velocity, pressure, and oxygen concentration) and all material properties (such as density and viscosity) are stored at the cell center. Boundary conditions of fixed value or zero flux are assigned through the use of fictitious boundary nodes for computational cells adjacent to the boundary. A variety of solution schemes are available. First-order upwind differencing in space and the Euler method in time were found to give good results for this problem. At each timestep, the solution procedure is repeated until either a specified convergence property is obtained or until a maximum number of iterations is reached. The iterative method used for this problem was the Conjugate Gradient Squared (CGS) + Preconditioning equation solver.

The material properties of the gases used in this model were taken from the standard tables of gaseous species. The viscosity for the mixture of gases was calculated using Sutherland's Law and the mix kinetic theory of gases. The ideal gas law was used to relate density, pressure, and temperature. This model did not take heat transfer into account. So although in reality the mix of fuel and air in the exterior environment will be at elevated temperatures, the material properties used in the analysis are room temperature values. Note that, in practice, the temperatures of the gases cannot be so high that they prohibit the presence of the firefighter.

Gravity was neglected in this problem.

4.2 Chemistry

The Chemistry Module in CFD-ACE+ enables the solution of mixing and reacting flow problems. In this model, oxygen from the leak in the respirator facepiece mixes with the fuel-air mixture in the surrounding environment. So the concentrations of each chemical species must be determined as they vary in time and space. To follow each of four chemical species (C_3H_8-propane, O_2-oxygen, N_2-nitrogen, and Ar-argon) separately, the Species Mass Fractions option was selected. (Although argon was included in the problem, it makes up a small fraction of air and could have been neglected.)

CFD-ACE+ contains a chemistry database that holds basic data for many elements, species, and mixtures.[17] A new mixture may be added to this database by specifying the mass fraction or concentration of each species in the mixture and assigning a unique name. The new mixture is then available by name for use for setting boundary and/or initial conditions.

Because oxygen is present in this problem both as an individual gas and as a component of the fuel-air mixture, the Species Mass Fraction option is chosen. In this approach, a transport equation is solved for every species in the problem. The mixing of oxygen from the leak with the fuel-air mixture requires a model for mass diffusion. A constant Schmidt number model is assumed, in which the mass diffusivity is assumed to be equal to the viscosity divided by a constant, the Schmidt number. The Schmidt number is assumed to equal 0.7 for this model.

Some care needs to be taken in interpreting the gas concentration values in the analysis results, since the variables in the CFD-ACE+ calculations are mass fractions and the lower and upper flammable limits (LFL and UFL) are volume fractions. To plot the contours of LFL and UFL, therefore, the mass fraction of propane under these conditions needs to be calculated. This is accomplished by using the molecular weights of each species in the mixture. For example, the UFL of propane in air at standard temperature and pressure (STP) is 9.5 % by volume. Pure air is 78.1 % N_2, 20.9 % O_2, and 0.9 % Ar by volume. (Note that in the flammability diagram, argon may be added to nitrogen as an inert gas.) The concentrations by volume of the components of air in a UFL propane-air mixture are obtained by multiplying the values in pure air by (1.00-0.095) = 0.905 to get 9.5 % C_3H_8, 70.7 % N_2, 19.9 % O2, and 0.8 % Ar. The molecular weights for each species are:

C_3H_8:   MW = 3 (12.01) + 8 (1.01) = 44.1 g/mol
N_2:   MW = 2 (14.01) = 28.0 g/mol
O_2:   MW = 2 (16.00) = 32.0 g/mol
Ar: MW = 1 (39.95) = 40.0 g/mol

The conversion from volume fraction C_i to mass fraction Y_i for species i with molecular weight M_i is given by:

$$Y_i = \frac{C_i M_i}{\sum C_i M_i}$$

Therefore, the value of mass fraction of propane that corresponds to the UHL in air is

$$Y_{UHLair} = \frac{(0.095)(44.1)}{(0.095)(44.1)+(0.707)(28.0)+(0.199)(32.0)+(0.008)(40.0)} = 0.138$$

Similarly, the mass fraction of propane at the LFL in air of 2.1 % by volume is 0.032.

For the UHL and LFL of propane in pure oxygen, there are only two species present. For the UHL in oxygen of 58 %, the volume fraction of oxygen is 42 %. The mass fraction of propane at this level is

$$Y_{UHLO2} = \frac{(0.58)(44.1)}{(0.58)(44.1)+(0.42)(32.0)} = 0.656$$

Similarly, the mass fraction of propane at the LFL in pure oxygen of 2.1 % by volume is 0.029.

4.3 Turbulence

For fluid flow problems, the presence of turbulence may strongly affect the transport of mass, momentum, and energy. Generally speaking, turbulence should be taken into account if the Reynolds number Re in the problem exceeds 1000, where $Re = \rho UL/\mu$, ρ is density, U is a typical velocity in the problem, L is a length scale, and μ is viscosity. Turbulence may be a factor if the Reynolds number is above 100. In this problem, the largest velocity and smallest length scale are both found at the leak. For a normal breath, the velocity at the leak is about 1 m/s, as will be shown in section 4.5. The width of the leak gives a length scale of 1 mm. For oxygen at STP, the density is 1.3 kg/m^3 and the viscosity is 2.0×10^{-5} kg/m-s. The resulting value of Re is about 50, low enough that turbulence is not necessary for calculations under conditions of normal breathing. If the first responder is breathing more heavily under stress, the velocity at the leak will be significantly higher, and a turbulence model may be advised (although turbulence will still be a factor only close to the leak).

When the turbulence module is employed, the standard k-epsilon model is chosen.[16] This model assumes the eddy viscosity approximation, in which the Reynolds stress tensor is proportional to the rate of mean strain, and solves transport equations for the turbulent kinetic energy k and the rate of dissipation ε.

4.4 Problem Geometry

The solution space for the outward leak problem is a volume exterior to the combined headform and respirator facepiece. Because of the symmetry along the centerline of the face and mask, the size of the computational problem to be solved can be cut in half. It should be understood, however, that the assumption of geometrical symmetry also requires that the outward leak is mirrored on the opposite side. As discussed in the section on Model Geometry above, the headform and respirator facepiece were developed with symmetry in mind. For ease of creation, a rectangular box was chosen, with the headform / facepiece roughly centered along one side. Fig. 6 shows the geometry of the computational problem, using only the exterior surfaces of the combined headform and facepiece from Fig. 5. The face inside of the facepiece has been removed and the visor and facepiece connector region below the visor have been filled in.

The rectangular box must encompass a sufficient volume of the surroundings such that the locations of the walls of the box have a negligible effect on the results in the immediate vicinity of the leak. A study of sensitivity to the size of the box is discussed in section 5.1.

For solution of the flow field and gas species concentrations by the finite volume analysis of CFD-ACE+, the region enclosed by the exterior surfaces of head and facepiece plus the walls of the box is divided into a tetrahedral finite element mesh to be used for computing the solution. This is accomplished using CFD-GEOM to first divide the surface entities of the geometry into a 2D triangular surface grid, then to use the surface grid to divide the volume into a 3D tetrahedral mesh. The 2D mesh generation automatically concentrates the mesh in any area with fine features defined by the surface entities, including the region around the thin leak. Much larger cells are defined near the walls where fine detail is not necessary. Setting a transition factor to a reasonable value such as 1.1 limits the maximum percentage change in cell size from one cell to the next. Further control over mesh size is obtained by defining a maximum or minimum cell size, by defining a source point to concentrate small cells, and by using the edge entity to define surfaces. The edge entity can be split into a one-dimensional grid that specifies the locations for the points of the 2D triangular mesh, and is used to make the mesh finer in regions of interest. This approach was used both to improve the resolution of velocity across the thin leak, described below in section 5.1, and to refine the mesh in the direction of the jet, described in section 5.3.

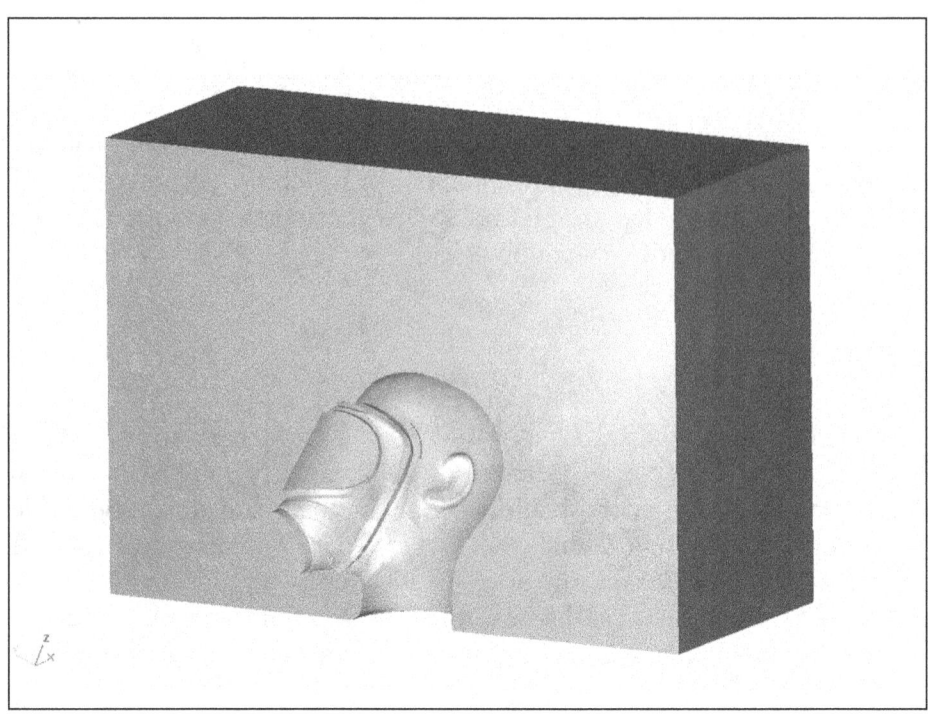

Figure 6. Model geometry for solving flow in the environment external to facepiece.

4.5 Boundary Conditions

To fully define the model, boundary conditions must be applied to all surfaces. There are four basic types of boundaries for this model: input, output, symmetric, and wall boundaries. Fig. 7 indicates where these boundary conditions are applied to the model.

As the critical element in this problem, the boundary conditions for the leak are of greatest interest. Because the flow is into the computational volume, this is an inlet boundary condition. Of the inlet flow boundary condition options available in CFD-ACE+, fixed normal velocity was selected as the most appropriate for this problem. The chemistry boundary condition is pure oxygen. Thus, at the leak oxygen flows into the volume with an assigned velocity profile. As a simplification of the breathing cycle, two sets of boundary conditions are used to represent exhalation and inhalation. During the exhalation half of the cycle, the leak is assumed to have a steady velocity. During inhalation, the leak is assumed to be cut off, with zero flow entering the computational space. In reality, an SCBA respirator is under positive pressure throughout the entire breathing cycle. The assumption made here is that the positive pressure is high enough to break the seal and cause a leak only during the exhalation part of the cycle.

As discussed in section 2.1, under normal rest conditions, a cycle of 15 breaths per minute with a tidal volume of 0.5 L is a reasonable assumption. Thus, over the two-second exhalation part of the cycle $0.5 \text{ L} = 0.5 \times 10^{-3} \text{ m}^3$ is released. Assume that a large fraction equaling 20 % of that breath is lost through the leak (and another 20 % is lost

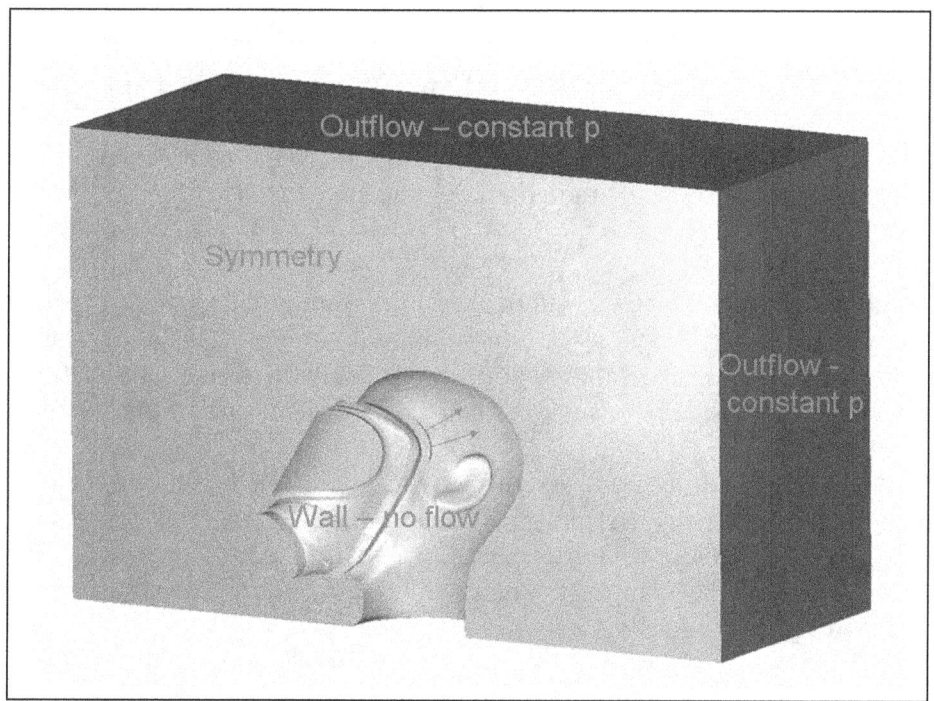

Figure 7. Model geometry illustrating boundary conditions

through the symmetric leak on the opposite side of the facepiece). Since the area of the leak is about 44 mm by 1 mm = 4.4×10^{-5} m^2, the velocity through the leak is approximately

$$v_{leak} = 0.20 \times \frac{\left(0.5 \times 10^{-3}\, m^3\right)}{\left(4.4 \times 10^{-5}\, m^2\right)\left(2\ s\right)} = 1.1\ m/s$$

The velocity through the leak for a normal resting breath is therefore taken as 1 m/s over a two second time period. For the following two seconds, the leak is closed. Then the leak of 1 m/s is open again for two seconds, followed by another two second closure, for a total of two breathing cycles.

To look at the effect of stress on the characteristics of flow from a leak, a case was run with an elevated breathing rate of 60 breaths per minute and tidal volume of 3 L. This corresponds to 0.5 s for each exhalation and inhalation half of the breathing cycle and a leak velocity of 6 m/s. Since the Reynolds number at the leak location is around 300, according to the discussion in section 4.3, turbulence is included in running this problem. For the inlet boundary condition, a standard guess of

$$k = 1.5\left(U \times TI\right)^2\ m^2/s^2$$

for turbulent kinetic energy and

17

$$\varepsilon = C_{\mu}^{0.75} \frac{k^{1.5}}{0.07\,L} \quad m^2/s^3$$

for turbulence dissipation rate are assigned, with $TI = 0.1$ the turbulence intensity, $C_{\mu} = 0.09$ a constant, $L = 1$ mm the characteristic dimension of leak width, and $U = 6$ m/s the reference velocity.

Outlet boundary conditions of fixed atmospheric pressure are assigned to the open sides, top, and bottom of the box defining the computational volume. This allows for flow both into and out of the model. The chemical content of any gaseous material flowing back into the volume from outside is assumed to be the fuel-air mixture of the surroundings.

Symmetric boundary conditions are assigned to the side of the box along the center plane of the combined headform and respirator. The conditions along this plane of symmetry are zero flux for all variables: velocity, species concentration, and pressure. This ensures that a mirror image of the computational volume about the plane of symmetry would yield a mirror image of the results.

The exterior surfaces of the combined headform and respirator, with the exception of the leak, are assigned a wall boundary condition. No flow is permitted through these surfaces.

4.6   Initial Conditions

At time t = 0, just before the leak due to exhalation begins, the computational volume is filled with the fuel-air mixture of the surrounding environment, and the flow is zero. Still air represents a worst case for oxygen buildup near the respirator, since any flow around the mask will act to mix the oxygen with the fuel-air mixture.

5 MESH STUDY

To make sure that the results are accurate and not dependent on the mesh, studies were carried out on the necessary resolution of the 1 mm wide leak, the mesh refinement in the direction of the jet, and the effects of the size of the computational box on the results. Since the time to solve the 3D problem is strongly dependent on the total number of grid cells, the goal is to keep the number of cells in the model as small as possible without introducing grid dependence.

A steady state problem was used to study mesh resolution and its effects on gas concentrations and velocities. The inlet boundary condition for this problem was a steady flow of oxygen through the leak at a velocity of 1 m/s. Initially the solution space was filled with propane. The calculation proceeded until a steady state condition of oxygen and propane concentrations in the region was reached.

5.1 Leak resolution

The leak region of this model is both the most challenging to resolve and the most important for the accuracy of the solution. The specified leak represents a thin breaking of the seal between the head and facepiece above the temple region during the exhalation part of the breathing cycle. The leak is 1 mm wide and 43.6 mm long and follows the curve of the facepiece seal along the headform. To properly set the boundary conditions here, the leak needs to be adequately resolved along the width.

One way to force a specific mesh resolution in a certain region using CFD-GEOM is to introduce edge entities, which are assigned a specified number of gridpoints along their length. The edges are then used to define the boundaries of surface entities. When the surfaces are meshed, the nodes of the mesh are forced to coincide with the gridpoints along the edges. The meshing algorithm tries to keep all sides of each triangle roughly the same, so the number of gridpoints along the length of the edge determines the average area of the 2D triangular elements in that region. A similar algorithm keeps the triangular sides of the 3D tetrahedral mesh in proportion, so the number of gridpoints also sets the average volume of the 3D tetrahedral cells adjacent to this boundary.

Fig. 8 shows the two-dimensional surface mesh along the leak resulting from three values of the number of gridpoints along each side of the leak. Above the plot of the 2D mesh for the entire leak is a close-up of the mesh near the upper endpoint of the leak. The purple dots are the gridpoints. The mesh in Fig. 8(a), with the edges along the sides of the leak divided into 87 gridpoints, provides a resolution across the leak width of only two triangles, as can be seen in the close-up. The resolution obtained by dividing each edge into 130 gridpoints in Fig. 8(b) is roughly three triangles across the leak width, and for 174 gridpoints in Fig. 8(c) is roughly four triangles.

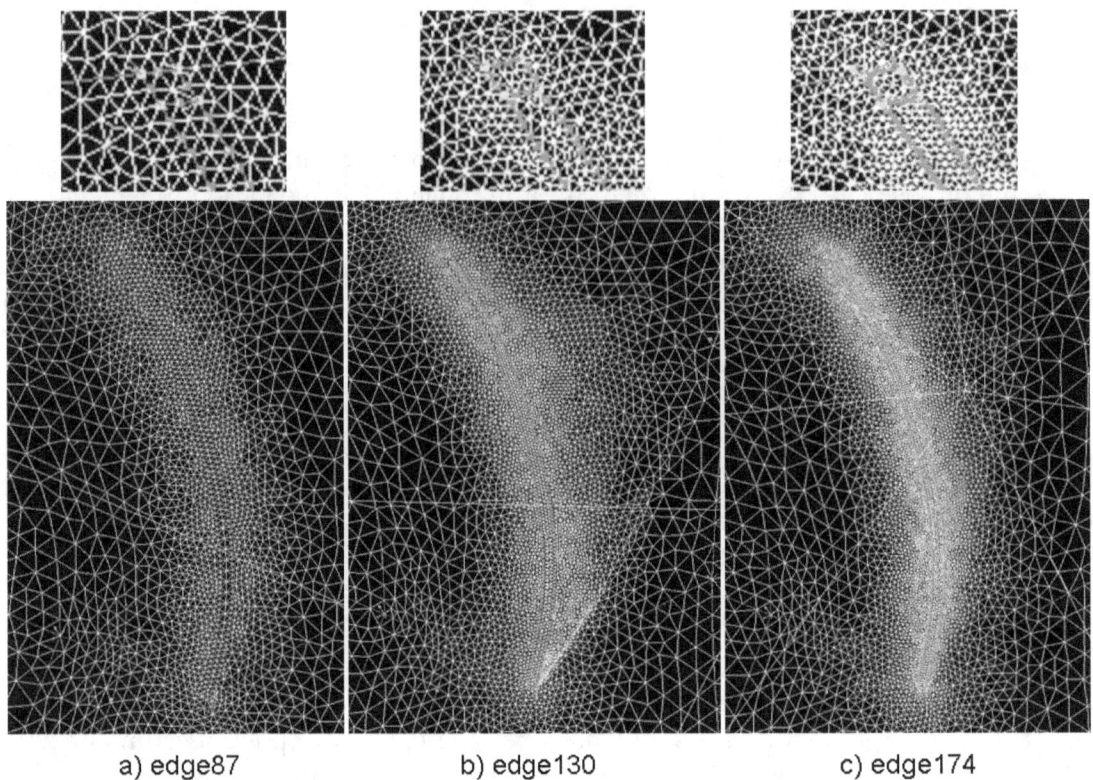

a) edge87   b) edge130   c) edge174

Figure 8. 2D mesh for three leak resolutions: a) 87 gridpoints along curve, b) 130 gridpoints, and c) 174 gridpoints.

Fig. 9 shows the profile of mass fractions across the leak for each of the three leak resolutions. These plots were obtained using the post-processing software CFD-VIEW, which provides tools for visualizing the results of CFD-ACE+ calculations.[18] To the right of each plot is a set of curves that delineates the surfaces in the region near the leak. The short line bounded by an x in each case lies on the model surface across the width of the leak. The plot shows the value of propane mass fraction along this line. Because oxygen and propane are the only gases in this model, the mass fraction of oxygen is equal to one minus the propane mass fraction. It is clear from Fig. 9(a) that the case using 87 gridpoints does not adequately resolve the mass fraction across the leak. However, both Figs. 9(b) and (c) show a smooth profile.

To make sure that the mass flow was correct, CFD-VIEW was asked to provide a boundary-by-boundary mass flow summary showing the inflow and outflow for each inlet and outlet boundary. These values were compared to the mass flux of oxygen flowing with a velocity of 1 m/s through a rectangle 1 mm wide by 43.6 mm long. The leak is a curved area rather than a rectangle, but no method was found for calculating the exact 2D area of the leak. The mass flow calculated by CFD-VIEW was within 2 % of the estimate for flow through a rectangle in each case.

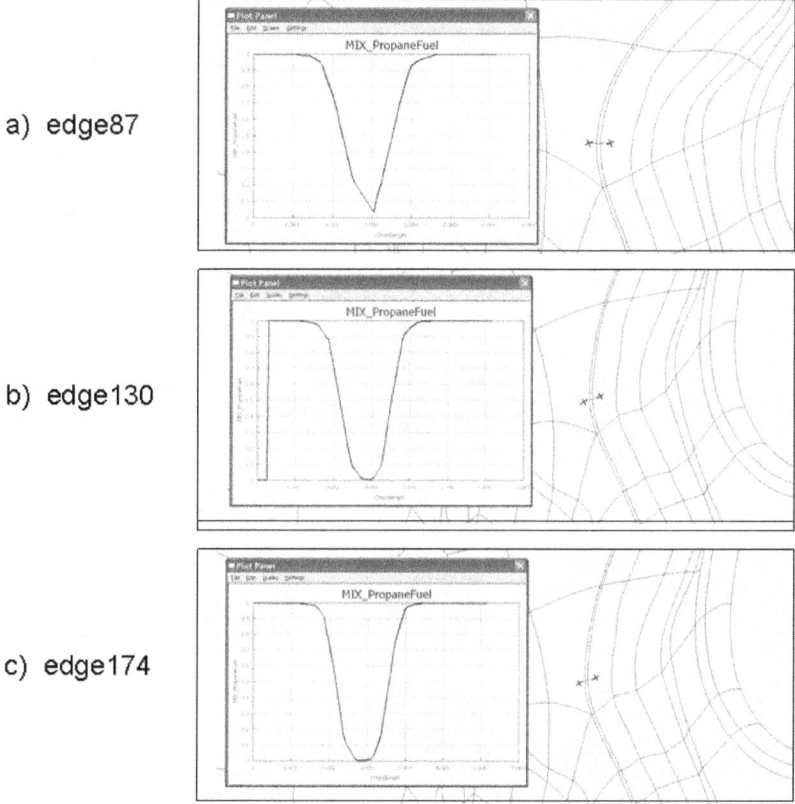

a) edge87

b) edge130

c) edge174

Figure 9. Comparison of propane concentration profiles for three leak resolutions

As the largest resolution (smallest number of gridpoints) of the three tested that gave good results, the number of gridpoints along the leak length was set equal to 130 for the final mesh. Fig. 10 demonstrates the location of the leak by drawing velocity vectors on a set of planes from just below to just above the leak.

5.2 Size of External Box

As discussed in section 4.4 and illustrated in Fig. 6, the computational space for this model was defined by a rectangular box into which the combined headform and respirator facepiece were embedded. For the first set of runs to test the model, the size of the box was chosen to be 600 mm wide in the direction to the right of the head, 900 mm long in the direction from front to back, and 600 mm high. A top view of this box is shown in Fig. 11(a). The shaded contour in this plot delineates the surface along which the mass fraction of propane is equal to 0.97 for the steady state problem described above. It was clear from the solutions of early problems that the size of this box resulted in a considerable amount of dead space over which gas concentrations and velocities varied little. A reduction of the size of the computational space to 600 mm wide by 600 mm long by 500 mm high reduced the number of grid cells and thus the analysis time without significantly affecting the results, as demonstrated by comparing 11(a) to 11(b).

21

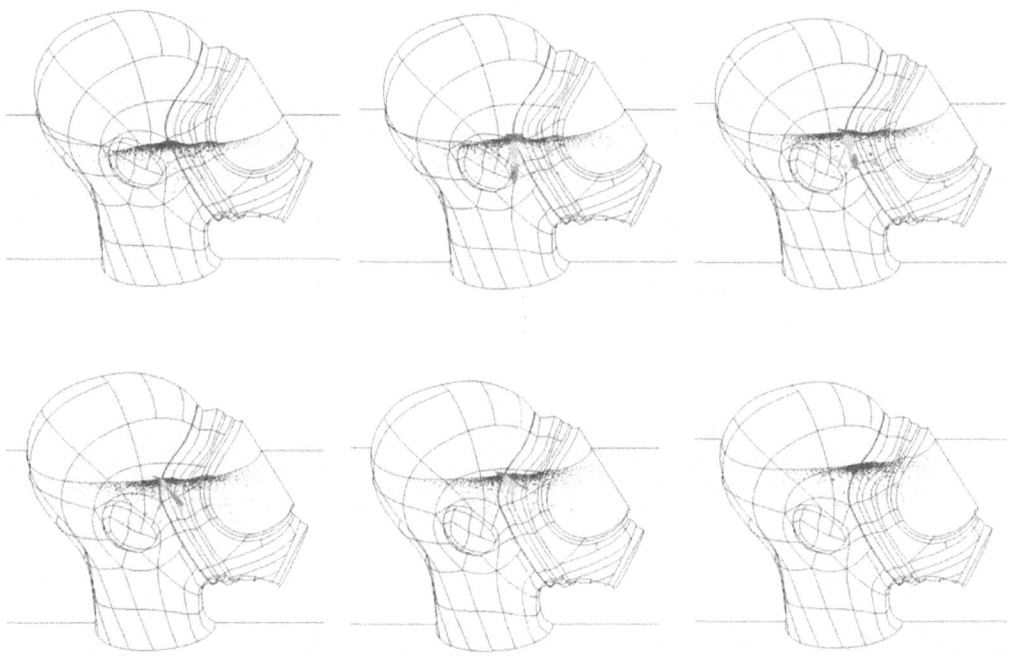

Figure 10. Velocity vectors on a set of planes across leak.

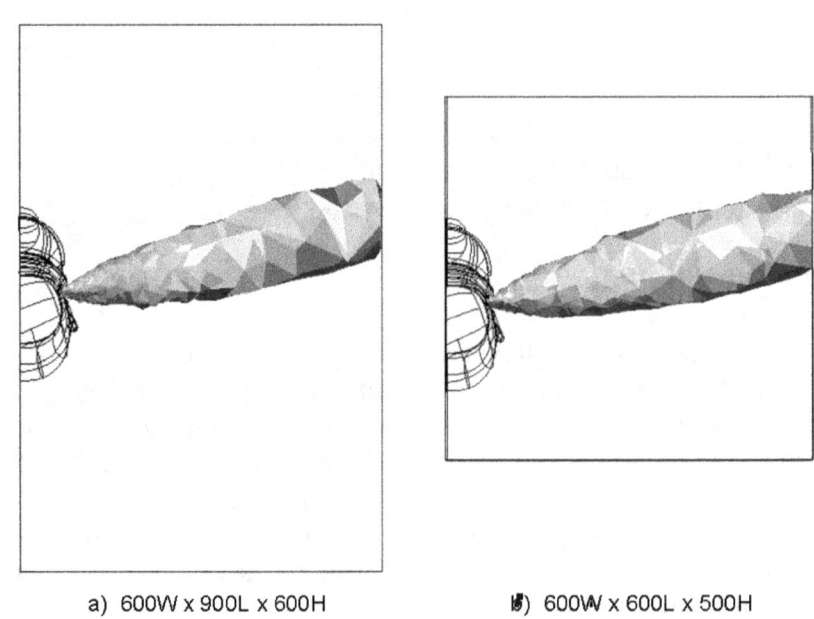

a) 600W x 900L x 600H   b) 600W x 600L x 500H

Figure 11 Comparison of propane mass fraction = 0.97 contour for two sets of dimensions for the computational space: a) 600 mm wide × 900 mm long × 600 mm high, and b) 600 mm wide × 600 mm long × 500 mm high.

5.3   Mesh Refinement in Direction of Jet

The oxygen from the leak flows into the solution space as a jet in the direction normal to the leak boundary. As shown in Fig. 11, this results in a region that extends in a more or less straight line from the leak to the side wall. The solution values within this region are of greatest interest. The previous section discussed how taking account of this region reduced the size of the computational space. It would be also advantageous to concentrate the mesh within this region in preference to the surrounding volumes. This can be accomplished using CFD-GEOM.

The first step is to locate the region of interest. Fig. 12 shows a view of the 0.97 contour surface from the side away from the head and mask, along with the same top view shown in Figure 11(a) (rotated by 90°). The post-processor CFD-VIEW is used to obtain plots of the mixture fraction of propane in the vertical z-direction (top plot) and horizontal y-direction (bottom plot), with the axes centered where the contour surface meets the side wall. This locates the centerpoint for generating a denser mesh.

The aim is to develop a 2D triangular mesh along the side wall that concentrates cells around the centerpoint. From this, the generation of the 3D mesh will fill in smaller grid cells through the entire region between the head and the opposite side wall, providing the desired improvement in resolution where it is needed. Two methods available in CFD-GEOM were used to produce a localized concentration of 2D grid cells: a point source and a set of edge elements in a circle. With the source method, the location of the centerpoint was identified along with a radius value that gives the length of cells emanating from this point. With the edge elements, a grid along each edge specified the desired resolution, in the same way as described for the leak mesh in section 5.1.

Fig. 13 shows the 2D mesh along the boundaries of the computational domain in the absence of mesh refinement along the side wall to the right of the head and facepiece. The grid along this wall, appearing in the lower left corner of the figure, is uniform and coarse. In Fig. 14, the source method has been used to obtain a higher density of cells in the region around the centerpoint of the jet. Fig. 15 shows the 2D mesh obtained from the use of edge elements arranged in a circle around the centerpoint. This grid was the finest and best controlled, and was selected to generate the final 3D mesh for the problem.

The number of grid cells in the final model was slightly under 500 000. Problems were run on a desktop workstation, with run times of less than 24 h for each exhalation or inhalation cycle.

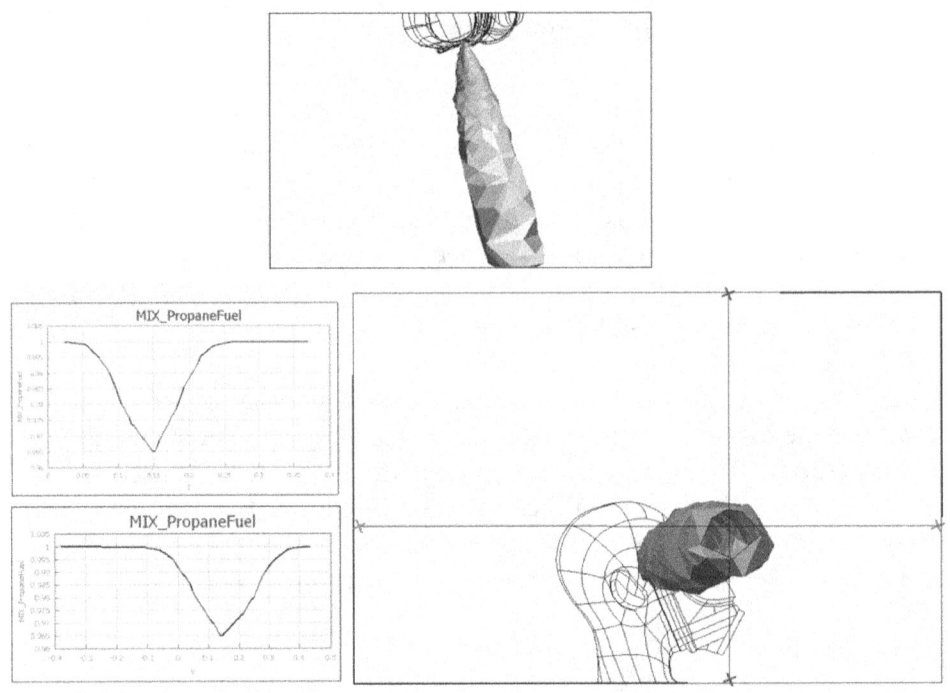

Figure 12. **Finding the point of minimum propane mass fraction along the side wall of the computational space. The 0.97 contour is shown for both top and side views.**

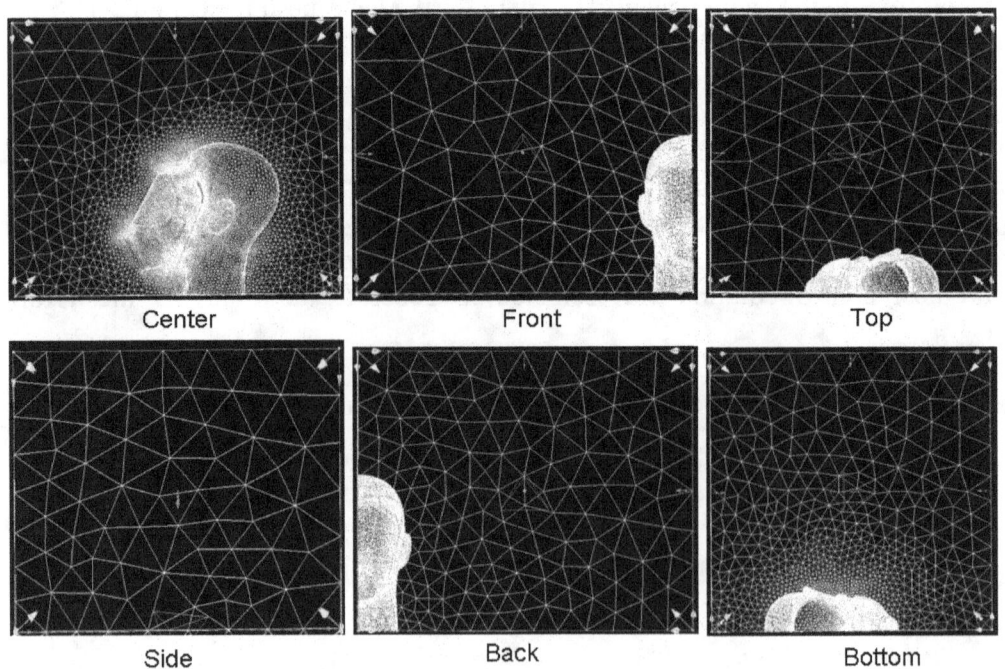

Figure 13. Mesh generation along boundaries of model space.

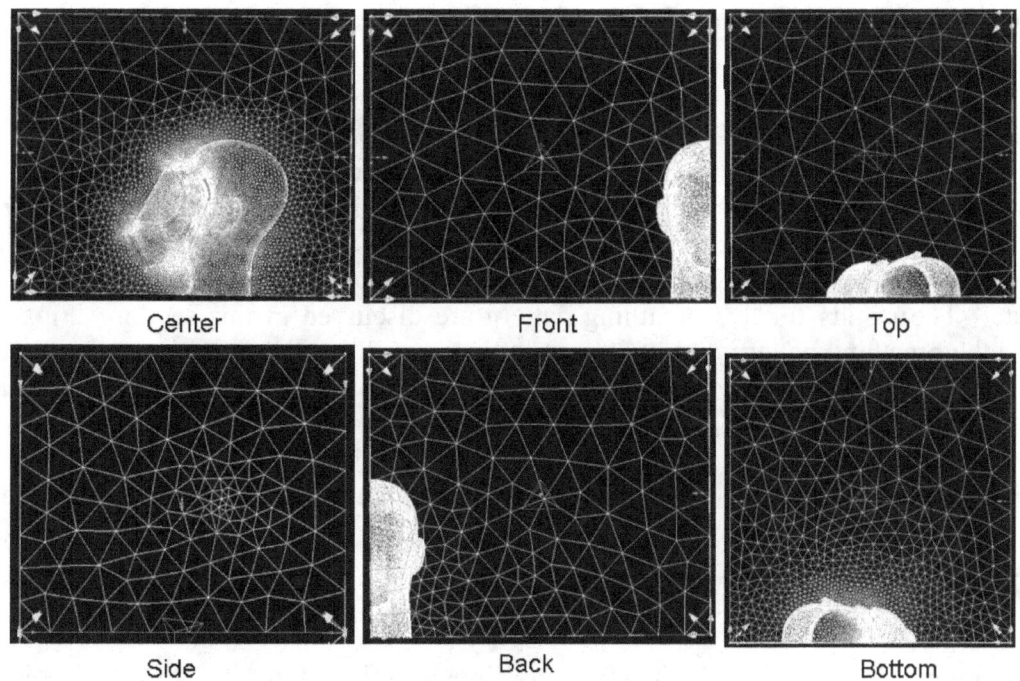

Figure 14. Mesh generated with source at side wall.

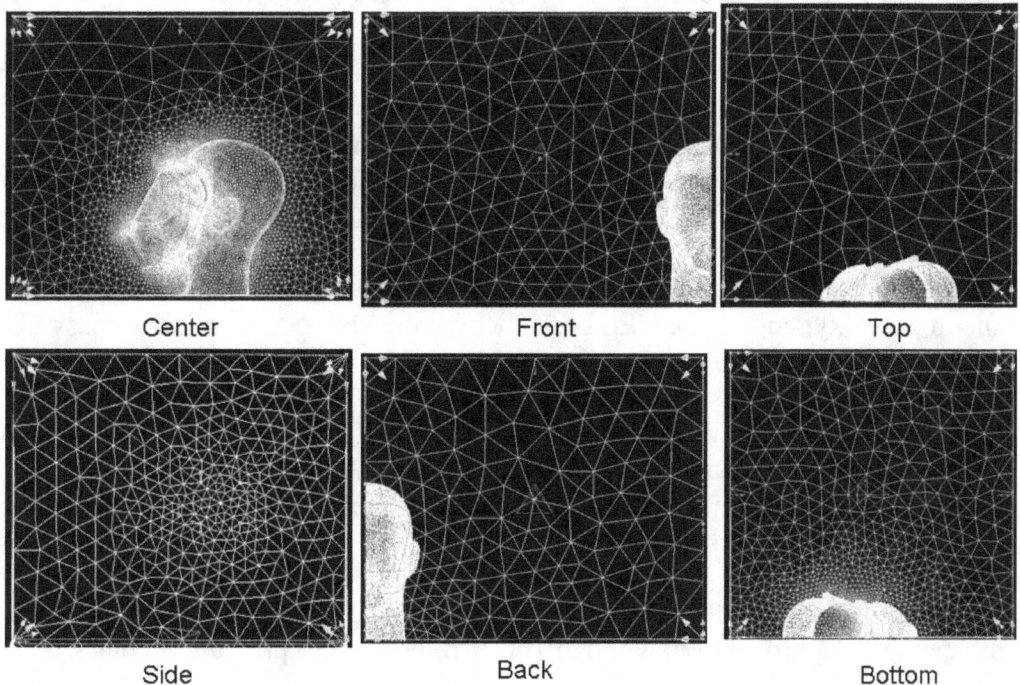

Figure 15. Mesh generated with circular edge element at side wall.

6 RESULTS FOR A SMALL LEAK DURING BREATHING AT REST

The effects of a respirator mask leak on the flammability of the nearby gas mixture depend on the oxygen content in the respirator and the fuel and oxygen content in the surrounding environment. Analyses were performed on three fuel-air mixtures, two levels of oxygen concentrations within the respirator, and two breathing patterns. Fig. 16 sketches the breathing patterns used in this study. Fig. 16(a) shows the breathing pattern for a person at rest, with tidal volume of 0.5 L and breathing rate of 15 breaths per minute. The results for this breathing pattern are discussed in this section. Fig. 16(b) shows the breathing pattern under high stress, which will be discussed in section 7. Both patterns assume a steady velocity from the leak during exhalation and zero leakage during inhalation.

CFD-VIEW provides several tools to visualize the three-dimensional flow field and gas concentration as functions of space and time. Results are displayed here on a horizontal plane located at the approximate midpoint of the leak, as indicated by the red line in Fig. 17. The view used for the display plane is from the top.

The flammability of the gases in the immediate vicinity of the head and facepiece is assessed by comparing the gas mixture at every point to the flammability diagram. If the mixture falls within the flammability region on the diagram, it is considered flammable.

6.1 Leak of Pure Oxygen into Pure Propane Environment

The first problem to be considered is pure oxygen leaking into a pure propane environment. In this case, the concentration of nitrogen is zero everywhere. Referring to the flammability diagram in Fig. 18, the mixtures in this problem all lie along the gold line on the left boundary, with values from pure oxygen at the lower left corner to pure propane at the top.

The flammable mixtures along this line lie between the LFL of 2.1 % by volume for propane in pure oxygen and the UFL of 58 % by volume. A conversion to mass fractions was performed as described in section 4.2 in order to correctly define the volume containing flammable gases in the results from CFD-ACE+. Fig. 19 shows a closeup of the area near the head at the end of the exhalation part of the breathing cycle. The red contour that extends outward from the leak by approximately a centimeter is the UFL contour. The LFL contour is very close to the leak. The area between these two contours is the flammable region in this problem.

The gray shading in Fig. 19 indicates the oxygen concentration. The values range from 0 % oxygen (100 % propane) in white to 100 % oxygen (0 % propane) in black.

Note that with both pure oxygen and pure propane in the problem, it is necessary to have a flammable region somewhere. This would also be true if the respirator was leaking

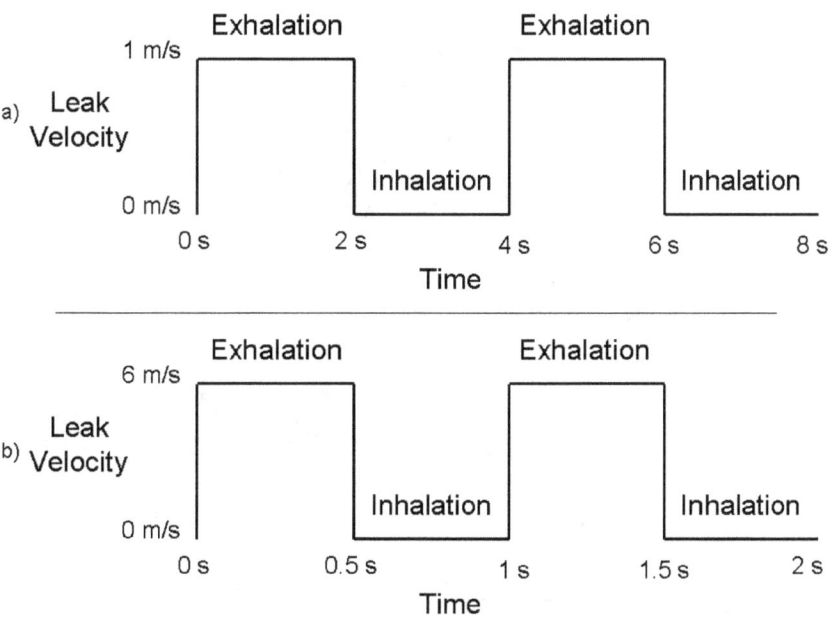

Figure 16. Breathing cycles for a) person at rest, and b) person under high stress.

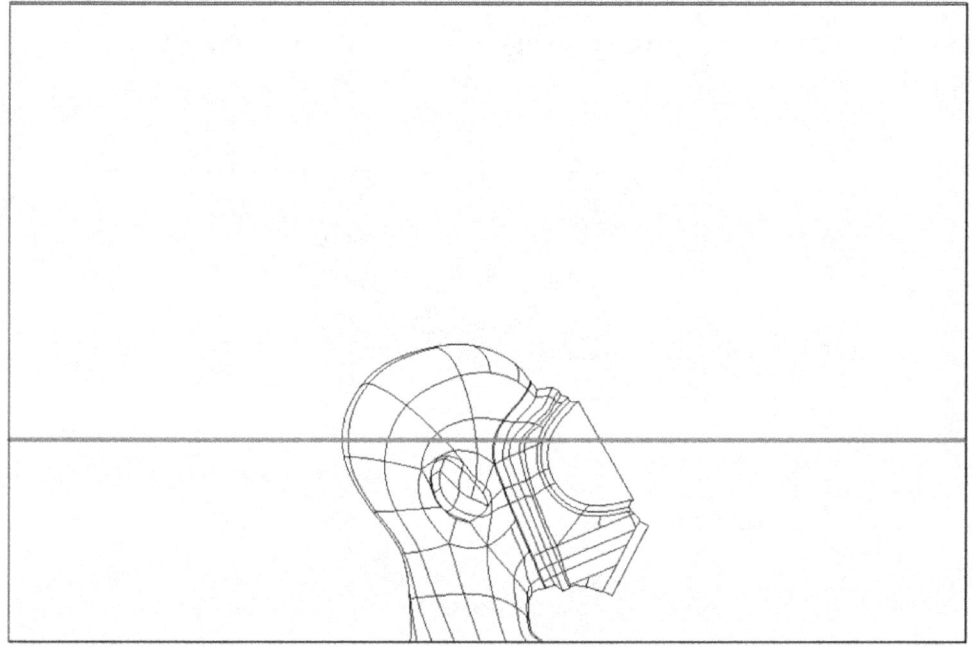

Figure 17. Vertical location of display plane.

27

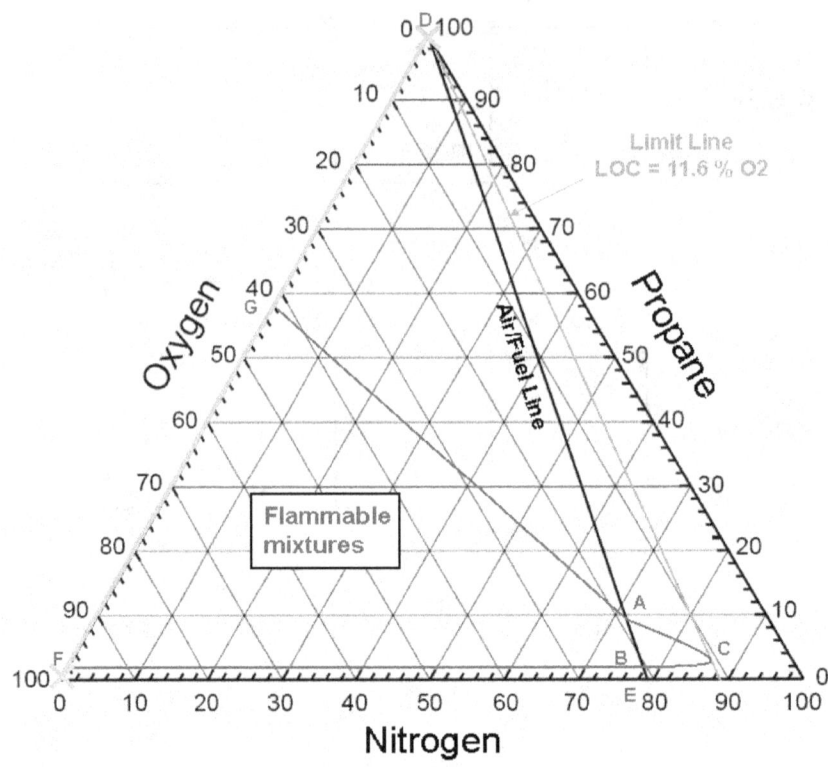

Figure 18. Flammability diagram showing range of mixtures for a pure oxygen leak into pure propane.

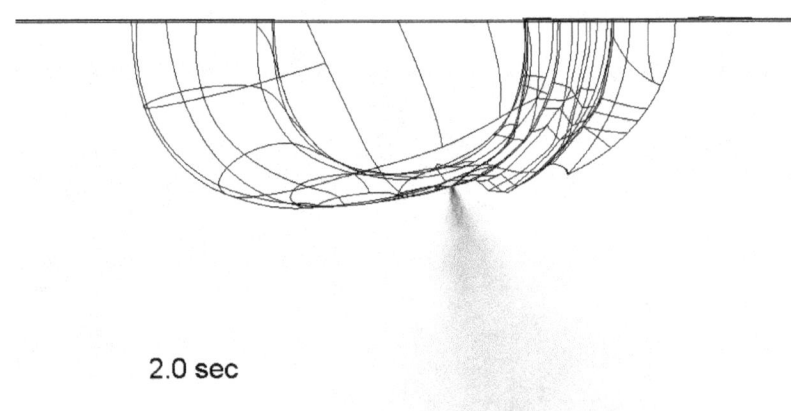

2.0 sec

Figure 19. Top view of head and respirator showing oxygen expelled during exhalation. Red contours mark the UFL and LFL of propane in pure oxygen.

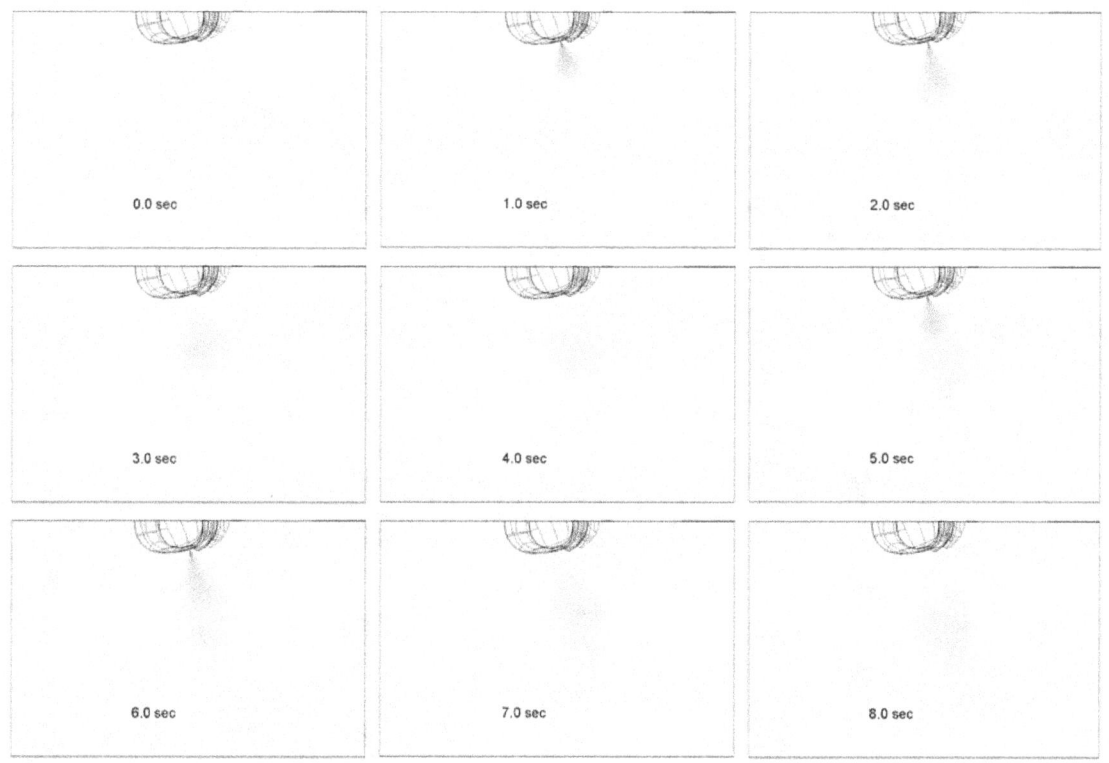

Figure 20. Time sequence of oxygen concentration from pure oxygen leak into pure propane environment.

either air or the 16 % by volume oxygen content of an exhaled breath. In the case of air, the gas mixtures in the problem follow the air/fuel line in the flammability diagram, and the flammable region is between the UFL and LFL in air. In the case of exhaled breath, the mixtures follow a line drawn between pure propane at the top and 16 % O_2 / 84 % N_2 along the bottom axis, which still crosses into and out of the flammability region in the diagram.

In this case, the flammable region near the head of the person wearing the respirator is small during exhalation and disappears completely during inhalation. Fig. 20 shows a time sequence of gas concentrations for these two breathing cycles. The evolution of the oxygen concentration is due to convection and diffusion.

Fig. 21 shows the flow field over time. During exhalation, between 0 s and 2 s and again between 4 s and 6 s, the flow at the leak itself has a value of 1 m/s, colored in red at the top of the scale. However, within a very short distance from the leak the velocity has decreased to a small fraction of that magnitude, less than 0.2 m/s. The jet from the leak creates a pair of rotating vortices, one on each side of the leak.

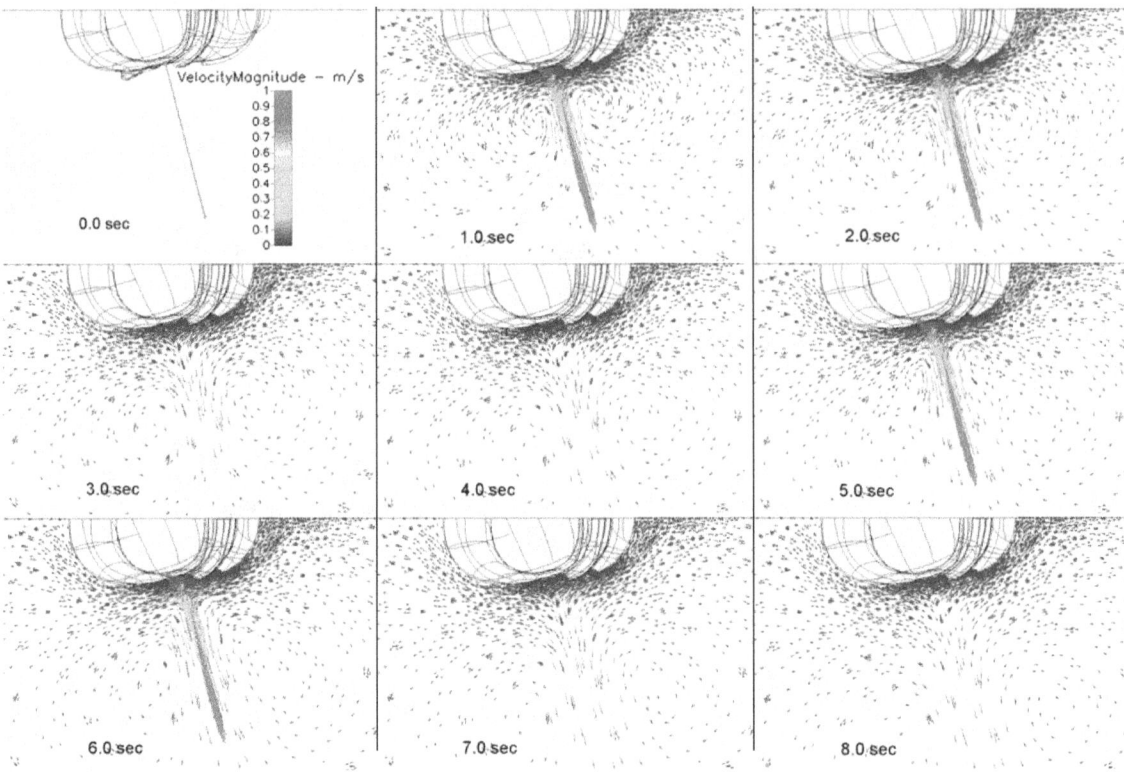

Figure 21. Time sequence of velocity vectors from pure oxygen leak into pure propane environment.

6.2 Leak of Pure Oxygen into 10 % Propane Environment

A much more challenging condition is the leak of pure oxygen into an environment in which the fuel level is very close to the UFL already, but just slightly too fuel rich for flammability. Since the UFL of propane in air is 9.5 % by volume, the value of 10 % by volume is assumed for the fuel/air mixture for this test. The introduction of oxygen into this environment will certainly result in a flammable region – the only question is what the size will be.

The flammability diagram is again used to determine the concentrations that bound the flammable region. In this case, all mixtures in the problem must lie along a line drawn from the 100 % oxygen content expelled from the leak to the 10 % propane point along the air/fuel line. This line is shown in gold in Fig. 22. The line crosses the lower curve bounding the flammable mixtures at 2.1 % propane, about 13 % N_2, and about 85 % O_2. At the upper curve, the mixture must be between the 9.5 % UFL value and the 10 % value of the surrounding mixture. It is important to get a good estimate for this mixture in order to get a good approximation for the size of the flammable region caused by the respirator leak. The mixture content is calculated as the intersection point between the

30

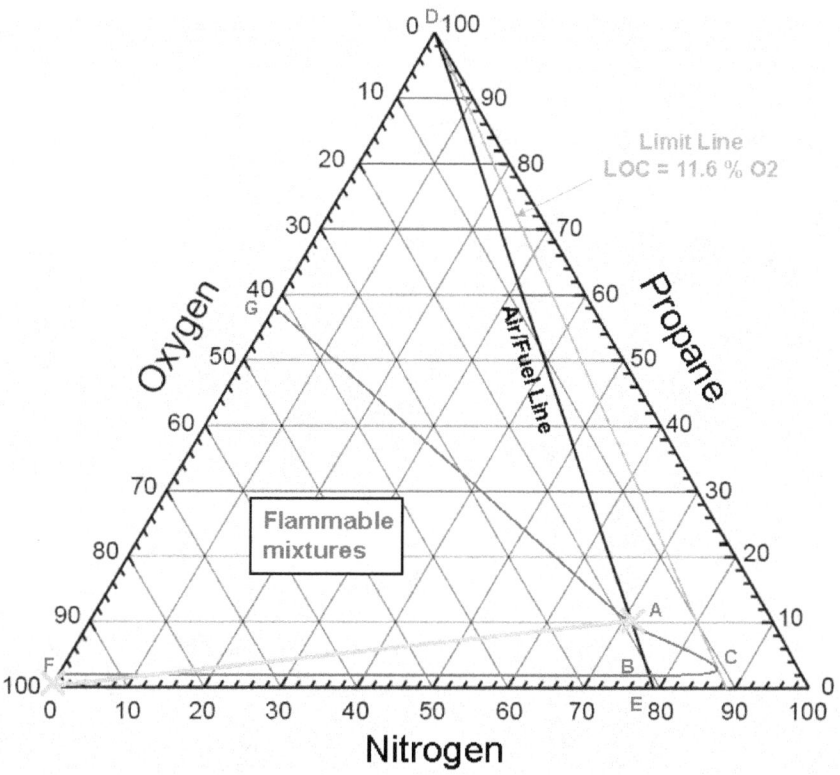

Figure 22. Flammability diagram showing range of mixtures for a pure oxygen leak into a 10 % propane environment.

line describing possible mixtures for this problem and the line between point A, the UFL in air, and point G, the UFL in oxygen. The propane concentration at this intersection point is found to be 9.96 % by volume.

Fig. 23 shows the time sequence for oxygen flowing into the 10 % propane/air mixture. In this case, the LFL contour is still very close to the leak, but the UFL contour has expanded considerably, indicating that there is a large region near the head that contains a flammable mixture. This contour is connected to the head throughout the breathing cycle.

Note that the growth in size of the flammable region over these two breaths is due to the zero leakage before time t = 0, a necessary assumption for starting this transient problem. The flammable region should eventually arrive at an equilibrium size (with some fluctuation over the breathing cycle) in which inflow from the leak is balanced by diffusion.

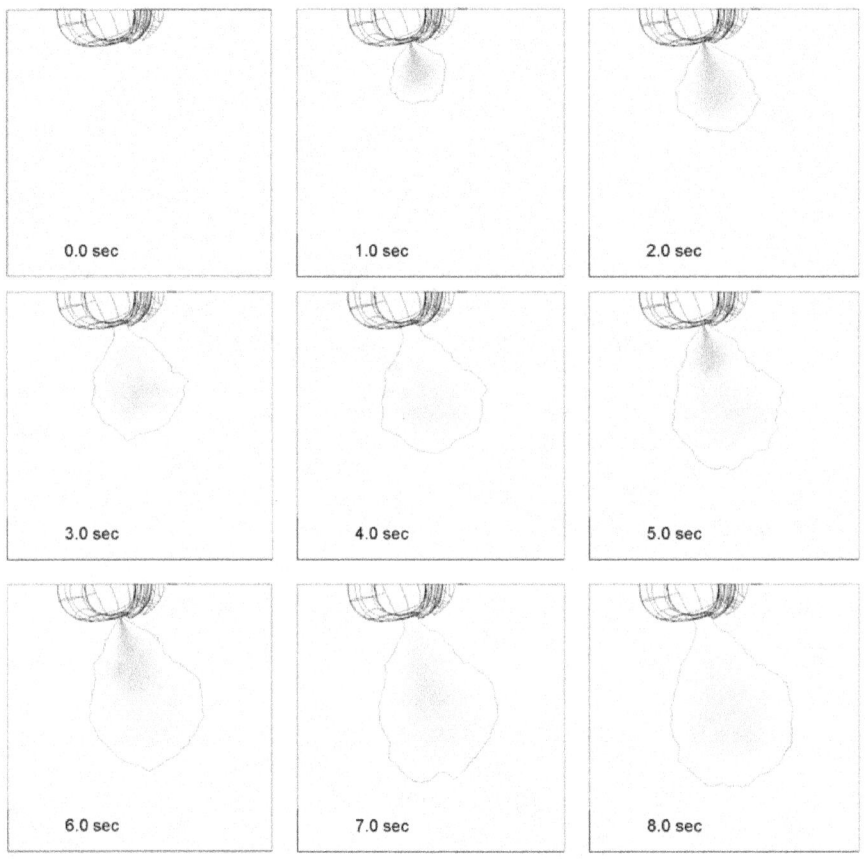

Figure 23. Time sequence of oxygen concentration from pure oxygen leak into 10 % propane environment, showing flammable region.

6.3   Leak of Air into 10 % Propane Environment

To appreciate the meaning of the result in the previous section for the use of closed-circuit SCBAs, it is useful to compare the leak from a respirator containing pure oxygen to a leak from a respirator containing air. The latter is a reasonable assumption for a standard open-circuit SCBA using a compressed air tank.

The gold line on the flammability diagram of Fig. 24 represents the range of mixtures in this problem, from pure air at point E to the 10 % propane point along the air/fuel line. Since this line lies along the air/fuel line, the points that bound the flammable region are already known to be the LFL and UFL of propane in air, 2.1 % and 9.5 % respectively.

Fig. 25 shows the time sequence of gas volume fractions for air leaking into a 10 % propane environment. Comparing this figure to Fig. 23 for a pure oxygen leak, it is clear that the flammable regions here are smaller. They detach from the head during inhalation and move away from the head and respirator. It is also clear that this environment is indeed one into which the introduction of oxygen from any source results in a flammable mixture.

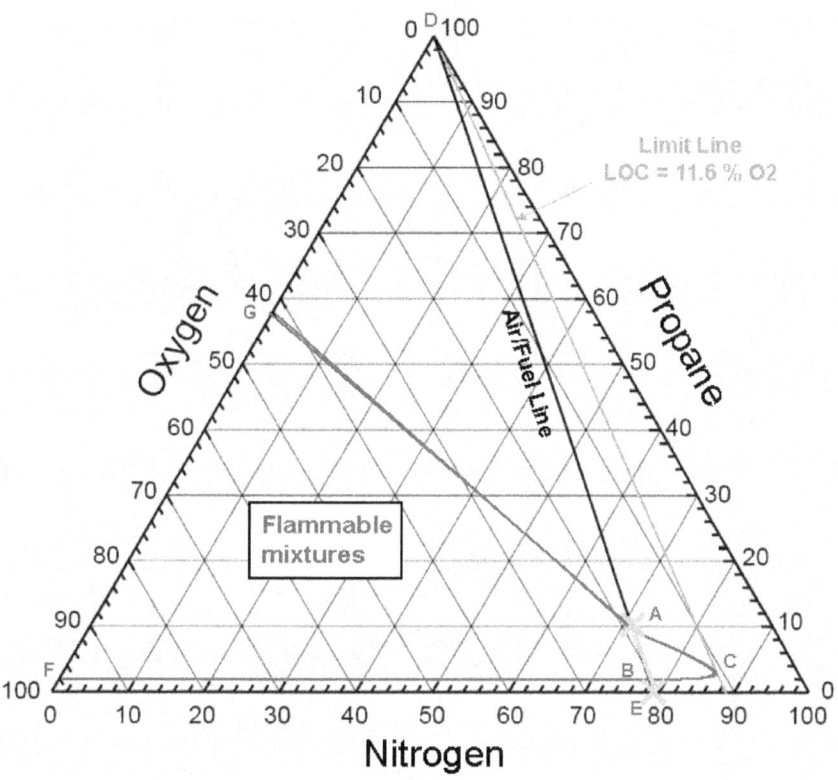

Figure 24. Flammability diagram showing range of mixtures for a leak of air into a 10 % propane environment.

6.4   Leak of Pure Oxygen into 5 % Propane Environment

The final problem studied for a resting breath is that of a pure oxygen leak into a 5 % propane environment. This is an environment that is flammable even without the addition of more oxygen.

The range of mixtures for this problem is plotted as a gold line on the flammability diagram in Fig. 26. In this case, the possible mixtures lie along a line from 100 % oxygen to 5 % propane along the fuel/air line. The point at which the line of possible mixtures crosses into the flammable region is at 2.1 % propane, about 24 % N_2, and about 74 % O_2. This LFL point corresponds to a mass fraction for propane of 0.296.

Fig. 27 shows a closeup of the area near the head at the end of the exhalation part of the first breathing cycle. The only contour in this plot is the LFL contour in red that extends outward from the leak by a few millimeters. The space within this contour is the only region in this case that is <u>not</u> flammable.

33

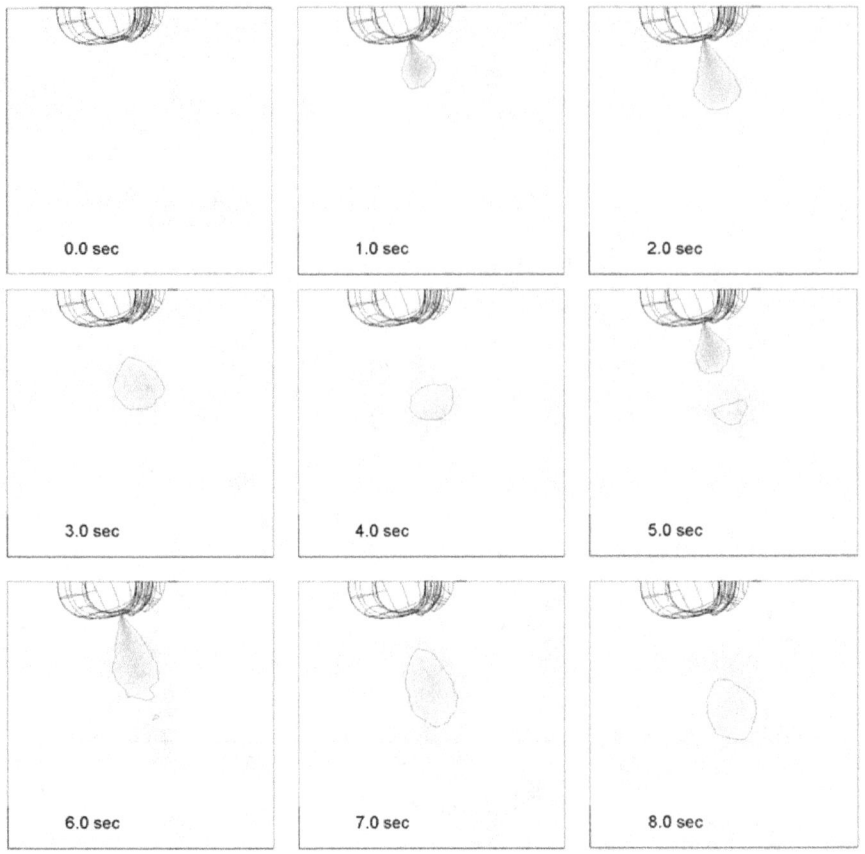

Figure 25. Time sequence of oxygen concentration from leak of air into 10 % propane environment, showing flammable region.

Fig. 28 shows the time sequence of oxygen concentration for oxygen flowing into the 5 % propane environment. The entire space is flammable, except for a small region next to the leak.

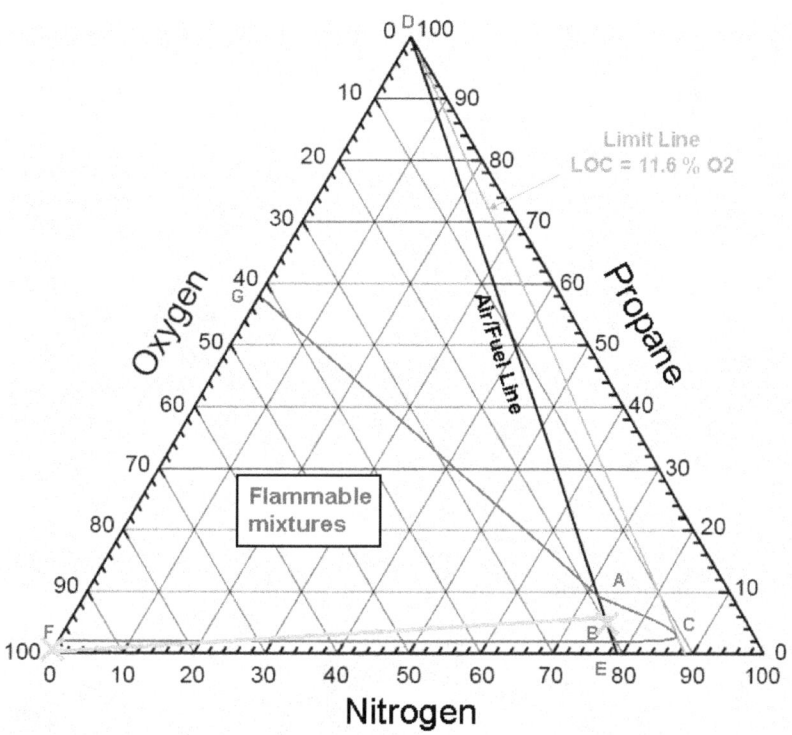

Figure 26. Flammability diagram showing range of mixtures for a pure oxygen leak into a 5 % propane environment.

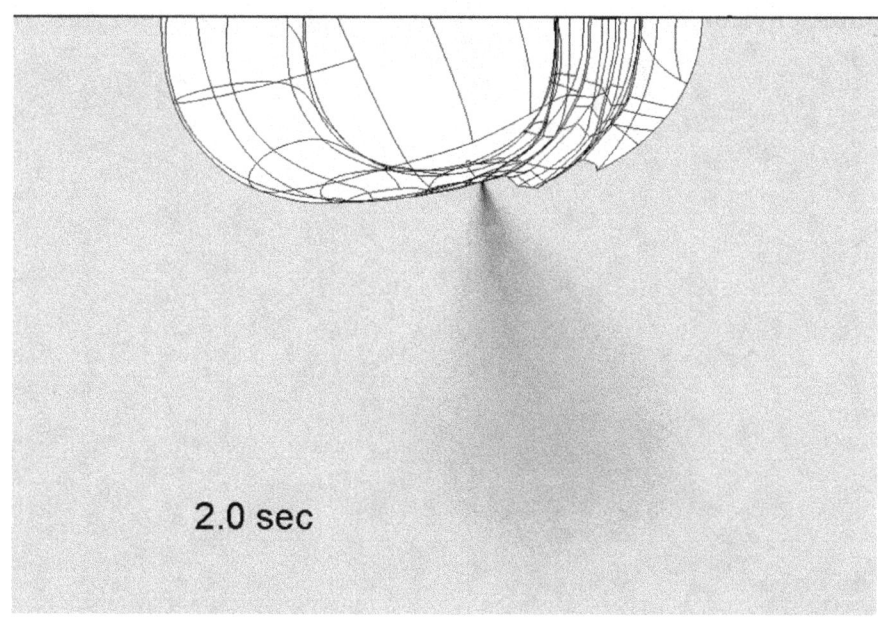

Figure 27. Pure oxygen expelled during exhalation into 5 % propane environment. Red contour marks the LFL of propane in this set of mixtures.

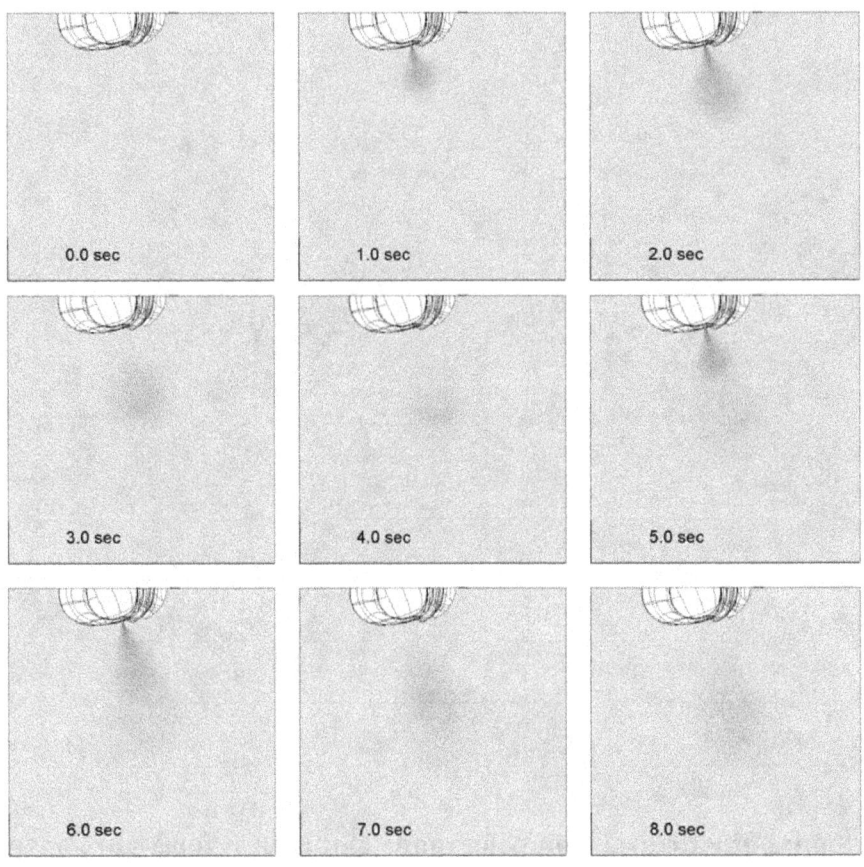

Figure 28. Time sequence of oxygen concentration from leak of pure oxygen into 5 % propane environment. (Note that the region <u>outside</u> of the red line is flammable.)

7 RESULTS FOR A SMALL LEAK WHILE BREATHING UNDER STRESS

For this final problem, the breathing pattern is changed to reflect the higher tidal volume and breathing rate for a person under stress from exertion. The breathing pattern is illustrated in Fig. 16(b), and represents a tidal volume of 3 L and breathing rate of 60 breaths per minute. The velocity boundary condition at the leak has been increased to 6 m/s, and a turbulence model has been added to the problem as described in section 4.3.

7.1 Leak of Pure Oxygen into 10 % Propane Environment

The most challenging environment is studied here, with pure oxygen leaking into a 10 % propane environment. These are the same conditions as in section 6.2, except for the higher leak velocity and breathing rate and the addition of the turbulence model. The set of possible mixtures is indicated on the flammability diagram of Fig. 22.

Fig. 29 shows the time sequence of oxygen concentration and UFL contours over the first two breaths. Comparing these results with Fig. 23 for breathing at rest, the spatial extent of the oxygen jet from the leak into the surroundings is longer. The oxygen disperses more quickly into the fuel/air mixture, and increases the size of the flammable region.

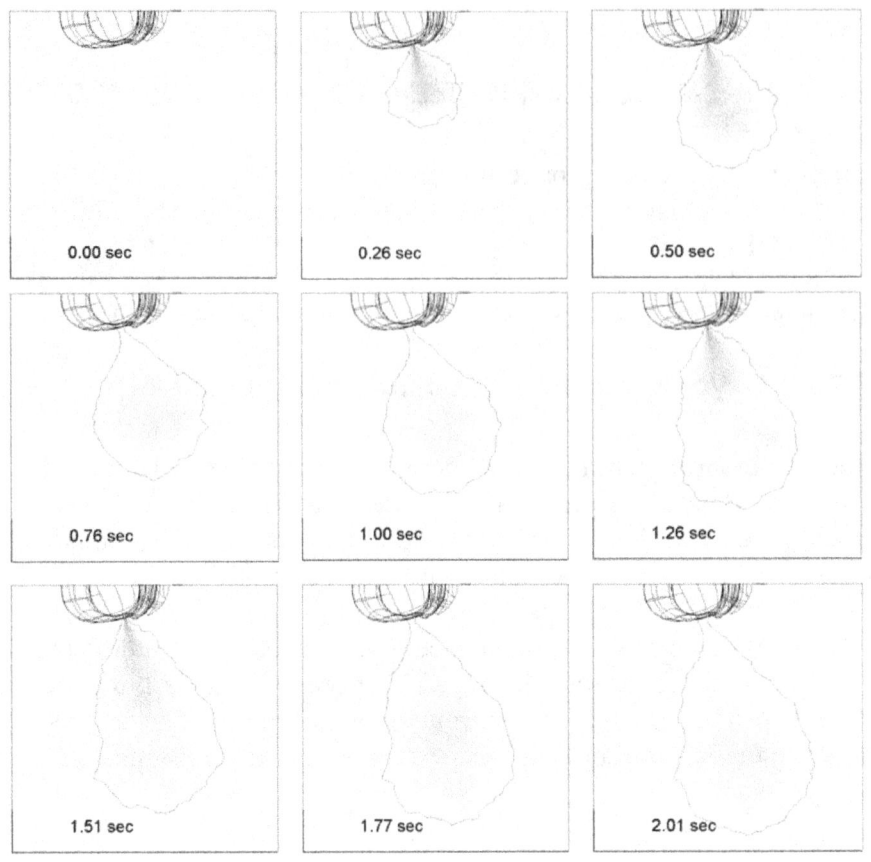

Figure 29. Time sequence of oxygen concentration from leak of pure oxygen into 10 % propane environment for breathing under stress, showing flammable region.

8   CONCLUDING REMARKS

This study has used computational fluid dynamics to investigate a respirator leak in a firefighting environment. The geometries of an experimental headform and a respirator mask were digitized and combined for an accurate model of the space surrounding the leak. Visualization tools allow the interpretation of scalar and vector data over time and space. This study demonstrates the capabilities of CFD to enhance our understanding of flow phenomena and communicate the results.

The focus of this study is on the potential dangers of a leak from the respirator of a closed circuit SCBA. Closed circuit SCBAs add pure oxygen to the breathing gases, which are scrubbed to eliminate carbon dioxide buildup. This compares to the open circuit SCBAs in widespread use, which supply air from compressed air tanks to firefighter respirators.

The worst-case scenario for a leak from either respirator is that of a slightly fuel-rich mixture in the surrounding environment, such that the introduction of oxygen at even low levels may result in flammability. The flammable region generated by a leak of pure oxygen is larger than that generated by a leak of air. Heavier and more rapid breathing, such as that caused by exertion, increases the flammable region further.

Under conditions of pure oxygen flowing into pure fuel, the flammable region is very small and disappears when the leak is closed off. When oxygen leaks into a flammable environment, a small non-flammable region is created very close to the leak.

In this study, the leak is assumed to be pure oxygen, rather than a lower concentration expected in the respirator by the combination of oxygen from the oxygen tank and breathing gases. The mixture of fuel gas and air in the worst-case environment is just above the upper flammable limit, a severe environment for a firefighter to be in. The surroundings are assumed to be still air, neglecting the flows normally associated with a fire and firefighter movement that will tend to mix the leak gases with the surroundings more rapidly. These flows will lessen the size of the flammable zone.

Validation of CFD with experimental measurements will ensure that these results accurately reflect reality. Of equal or greater importance is a better understanding of the actual conditions that may be encountered in the field, including the potential leak size and location, flow rates, and external environment.

REFERENCES

---

[1] Kyriazi, N. "Proposal for Certification Tests and Standards for Closed-Circuit Breathing Apparatus," Information Circular 9449, U.S. Department of Health and Human Services (NIOSH), Pittsburgh, PA (1999).

[2] Kyriazi, N. "Performance Comparison of Rescue Breathing Apparatus," Report of Investigations 9650, U.S. Department of Health and Human Services (NIOSH), Pittsburgh, PA, Oct 1999.

[3] De Vries, H., "Mining for Answers," Fire Chief, 1 Jan 2002.

[4] NIOSH/NPPTL, NIOSH/NPPTL Public Meeting – Powered Air Purifying Respirator Standards Development (for Protection Against CBRN Agents), Pittsburgh, PA, December 15, 2004. http://www.cdc.gov/niosh/npptl/standardsdev/cbrn/papr/meetings/121504/ (2004).

[5] NIOSH, Draft: Guide to the Technical Use of Chemical, Biological, Radiological, and Nuclear (CBRN) Open Circuit, Pressure-Demand Self-Contained Breathing Apparatus (SCBA) Respirators Certified Under 42 CFD Part 84. http://www.cdc.gov/niosh/npptl/pdfs/CBRNSCBAJuly15.pdf (2005).

[6] Stengel, J.W. and Rodrigues, R., "Machine Testing of Self-Contained Breathing Apparatus at a High Work Rate Typical of Firefighting," Journal of the ISRP, **2** (4): 362-368 (1984).

[7] W. J. Williams, "Physiological Considerations for Respiratory Protective Devices, Advanced Personal Protective Equipment: Challenges in Protecting First Responders," Virginia Tech, Blacksburg, Va, 16-18 October 2005.

[8] "Guide to Fire Hazard Properties of Flammable Liquids, Gases, and Volatile Solids", NFPA 325, 1994 edition, Fire Protection Guide to Hazardous Materials, 11th Edition, National Fire Protection Association, Quincy, MA (1994).

[9] Beyler, C., "Flammability Limits of Premixed and Diffusion Flames", Chapter 7, SFPE Handbook of Fire Protection Engineering, Third Edition, P.J. DiNenno et al. (eds.), Society of Fire Protection Engineers, Quincy, MA (2002).

[10] "Standard on Explosion Prevention Systems", NFPA 69, 2002 edition, National Fire Codes: A Compilation of NFPA Codes, Standards, Recommended Practices, and Guides, Vol. 4, National Fire Protection Association, Quincy, MA (2005).

[11] "Recommended Practice on Materials, Equipment, and Systems Used in Oxygen-Enriched Atmospheres", NFPA 53, 2004 edition, National Fire Codes: A Compilation of NFPA Codes, Standards, Recommended Practices, and Guides, Vol. 14, National Fire Protection Association, Quincy, MA (2005).

[12] Fredenslund, A., Gmehling, J., Michelsen, M.L. Rasmussen, P., and Prausnitz, J.M., "Computerized Design of Multicomponent Distillation Columns Using the UNIFAC Group Contribution Method for Calculation of Activity Coefficients", Ind. Eng. Chem., Process Des. Dev., **16**(4):450-462 (1977).

[13] Mashuga, C.V. and Crowl, D.A., "Application of the Flammability Diagram for Evaluation of Fire and Explosion Hazards of Flammable Vapors," Process Safety Progress, **17**(3): 176-183 (1998).

[14] Anthony, T. R., Flynn, M. R. and Eisner, A., "Evaluation of Facial Features on Particle Inhalation," Annals of Occupational Hygiene, **49**(2): 179-193 (2005).

[15] CFD-GEOM V2004 User Manual, ESI-CFD Inc., Huntsville, AL, www.esi-group.com (2004).

[16] CFD-ACE+ V2004 Modules Manual, ESI-CFD Inc., Huntsville, AL, www.esi-group.com (2004).

[17] CFD-ACE+ V2004 User Manual, ESI-CFD Inc., Huntsville, AL, www.esi-group.com (2004).

[18] CFD-VIEW V2004 User Manual, ESI-CFD Inc., Huntsville, AL, www.esi-group.com (2004).